The Iris

The Iris

Brian Mathew

B T BATSFORD LTD LONDON

© Brian Mathew 1981
First published 1981

ISBN 0 7134 3390 6

Typeset by Deltatype Ltd, Ellesmere Port
and printed in Great Britain by
The Anchor Press Ltd,
Tiptree, Essex
for the publishers
B. T. Batsford Ltd
4 Fitzhardinge Street
London W1H 0AH

This book is
dedicated to the memory of
Paul Furse

Contents

Foreword

Originally I had planned to write on *Iris* and her relatives, to include all those other iridaceous plants around the world which have iris-shaped flowers, such as *Tigridia, Patersonia, Ferraria, Moraea* etc. Fortunately I began with *Iris* and by the time that section was completed it was quite clear that I had set myself a hopeless task, for reasons of space and time. The contents of the book are therefore mainly confined to the genus *Iris*, with a few other small related genera. I have included only those other genera which occur in northern temperate regions since most of the others from South Africa, South America, Australia and New Zealand are not hardy in Britain, nor much of North America for that matter. *Pardanthopsis* is incorporated because it looks just like an iris and has in the past been included in the genus. *Belamcanda* is also put in since it hybridizes with *Pardanthopsis* and is obviously very closely allied to *Iris*. *Gynandriris* occurs in both Mediterranean regions and South Africa and is included because one of its northern hemisphere species is very commonly seen, and is often referred to as *Iris sisyrinchium*. I have also written about the monotypic genus *Hermodactylus*, for it is Mediterranean and is frequently mentioned in literature as *Iris tuberosa*.

In order to provide some sort of logic to the book I have provided a key to these genera and have split *Iris* itself into Subgenera, Sections and Series, with keys to those also. My aim has been to provide a book which the iris enthusiast can use to find a description of almost any wild iris species, together with its geographical distribution and habitat. Coupled with this I have provided notes on cultivation and on the relationships between the various groups, and between the individual species in them.

I hope I have achieved something between a stuffy reference book and a chatty gardeners' guide. Unfortunately at the present it is impossible to produce a thorough botanical revision of the genus *Iris*, for so many of the areas where irises occur are out-of-bounds. It will be very many years—several decades at least I imagine—before such a work is possible. In the meantime, iris enthusiasts need all the reading matter they can find to pursue their hobby to the fullest extent.

Brian Mathew

The illustrations

Acknowledgements Acknowledging the help of friends and colleagues is always difficult, for there are so many aspects to be considered. Who does one thank the most? Those who have given me iris plants, those whose published works I have consulted, those who have collected herbarium specimens, those who have grown irises and given me the pleasure of seeing them in the living state or those who have passed on knowledge through informal 'chat'? Some people have taken me to see irises growing in their natural state and in other cases I have received photographs of plants in the wild, which is the next best thing to visiting them.

The truth is that one cannot really single out individuals and say that they have been of more help than others. The person who has given me one packet of seeds of some rare species has contributed as much as any other to my experience, and this book would be incomplete without it.

Throughout the text, I have made reference to people who have been involved in some way or other and I hope that no one will be offended if I offer my thanks in a general way to everyone who has been mentioned.

The line drawings have been prepared by Miss Pat Halliday and I am grateful to her for remaining enthusiastic through the seemingly unending incubation period of the book. To Maggie I apologize for the tetchy state in which I have been for the same period and can only say thank you for tolerating it, and for typing the end product.

The author and publishers would like to thank the following for their permission to reproduce the colour photographs in this book: T. Baytop (3); M. Boussard (14,31); P. Furse (8,22); A. Güner (28); F. N. Hepper (7); S. Walker (15); P. Wendelbo (26). The remaining colour photographs were taken by the author. All the black and white photographs are by Erich Pasche.

Preface

As far as possible, the information given in the following pages is from my own observations on living and herbarium material. Obviously I have had to refer to some literature, but I have tried to minimize this as far as possible to avoid excessive plagiarism.

The list of species names does not include all the synonyms in existence, for the task of thoroughly checking every name against its description and specimens would be enormous and beyond the scope of the book. However, I have tried at least to mention all the latinized specific names which an enthusiast is likely to encounter even if, for example, only to state that the name refers to a hybrid, or is a poorly described plant from western China and the only specimen has been lost!

I have included the names of the authors of each species — that is, the names of the botanists who first gave the species in question a correct Latin name and description. There is so much confusion of species names within the genus that one has to be precise in order to make the meaning clear. For example, *Iris orientalis* Miller is an entirely different plant from *Iris orientalis* Thunberg and without the author's name, confusion would reign supreme!

A few words of explanation about the naming of plants are therefore probably worthwhile for, to the non-botanist, the whole business may be somewhat mysterious.

In an attempt to standardize the method of plant naming throughout the world there is an internationally agreed set of Rules of Nomenclature. These rules are published in book form and are therefore available for study; since they are fairly complex I have no intention of going into great detail about them here, for the enthusiasts who require information such as this are few and far between and it is better for them to study the Rules at source.

Our present system of naming largely follows that of Linnaeus who in 1753 gave plants two names, a generic one (e.g. *Iris*) and a specific one (e.g. *spuria*), instead of the previous practice of supplying them with longer descriptive names. By placing the author's name after the specific epithet, for example *I. spuria* Linnaeus, we show that Linnaeus was the botanist who gave the plant a botanical description under that name. The date of 1753 is the 'turning point' in nomenclatural history, any name published before that being unacceptable.

The rule concerning priority, in essence, says that the oldest available name for a species (1753 or after) is the correct one. Any other, later, names referring to that species are said to be synonyms. Thus, as an example, for the species *I. orientalis* Miller (1768) there is a later name, *I. ochroleuca* Linn. (1771), which is therefore a synonym: the former is the correct one by reason of priority of date of publication. In some cases, one will find the names of two or more authors after the specific epithet and this may be so for various reasons: 'Boissier & Heldreich' indicates joint authorship; 'Boissier ex Heldreich' means that Heldreich published the description of a plant which had been referred to, but not validly published under the rules, by Boissier; '(Boissier) Heldreich' would inform one that Heldreich had changed the status of the plant in some way, for example moved it from one genus to another or changed its 'taxonomic level' (e.g. variety to species, forma to variety, etc.). More names can be involved with joint authorships, so that one might see something like '(Boissier & Heldreich ex Baker) Dykes & Lynch'.

Ideally for every species described (to be correctly published there must now be a Latin description) there is a 'type specimen' in existence. This is the actual plant which was used to make up the description and is normally in the form of a dried specimen preserved in a herbarium. This specimen sets the standard for the species upon which all future comparisons must be made. Unfortunately many of the older names were based on living material only, which was not preserved, or the type specimens have been subsequently lost or destroyed. In some of these cases, especially if the description is scanty, it is not always easy to be sure that the name is being applied correctly and in certain instances it may largely be a matter of opinion, one taxonomist's view against another's. Nowadays a type specimen must be preserved and it is preferable that this should be from wild rather than of cultivated origin, and accompanied by good field notes.

For reasons of space I have not included in this book the place of publication of each name. The reader may obtain this information by referring to the *Index Kewensis* or to the invaluable booklet by the late Peter Werckmeister, *Catalogus Iridis* (1967) which gives lists of most, if not all, the names published in the genus *Iris*, with their literature references and dates.

In order that the real addict may follow up his particular line of interest, I have also included at the end of the book a list of published works for suggested further reading. This list is of course far from complete, for there are countless small (but nevertheless important) papers on individual species in addition to all the major works. I have merely made a selection of the more substantial or important and these will inevitably lead the reader via the 'snowball effect' to endless other publications.

In general, I have not dealt with cultivars or even the lower taxonomic ranks, such as forma. Here again, space and time are the limiting factors. I have however mentioned the range of variability of each species as far as I can ascertain it. The only hybrids mentioned are those which have been given specific names, for example *I. squalens (I. variegata × I. pallida)*, since it appears in lists of species such as that given in *Index Kewensis*.

Cultivation notes are primarily based on my own experience in Surrey, or on information gleaned from literature. It is up to the reader to use or reject this knowledge as he thinks fit, for gardens even a few miles apart can vary enormously in their microclimates. There is no general chapter on cultivation since in a large genus such as this it would serve no useful purpose. The notes are therefore incorporated within each group of irises and, in special cases, in the accounts of individual species.

There is however, a general chapter on construction of iris beds, raised beds and bulb frames, and comments on pot cultivation.

Introduction

The iris plant

To describe the iris in a few lines is quite impossible, for there are a great many variations in this sizeable genus, from the roots and storage organs right up to the tips of the flowers. An attempt has been made with the help of line drawings to show the basic parts of some irises so that the reader will be more familiar with the terminology used in later parts of the book.

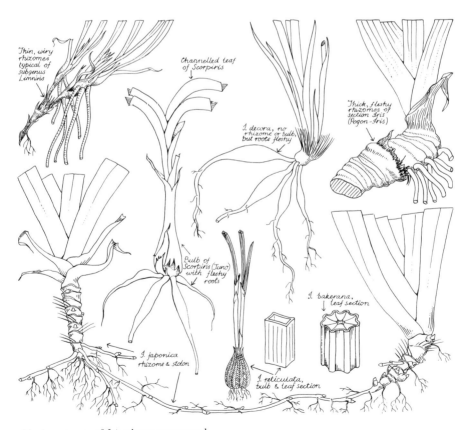

Various types of *Iris* plants compared

The roots and storage organs

The underground system of most irises consists of fibrous roots which may be few- to many-branched, thin and wiry or rather thick and fleshy. Usually there is also some form of swollen storage organ—most typically a fat rhizome, but there are many other types. They may have bulbs, covered with netted-fibrous, papery or leathery tunics. If they are rhizomatous, the rhizomes can vary from the familiar thick ones of the tall bearded irises to very thin and wiry ones, often far-creeping by means of stolons.

The leaves

There is a great range of leaf types throughout the genus, from flat to channelled or square to nearly cylindrical in cross-section. The flat leaves are usually widened at the base with a sheathing part, but the blade is carried in a vertical position, giving rise to the familiar flat fans of the common bearded irises. The channelled type often have a sharp keel on the underside. The longitudinal veins in the leaves may be scarcely noticeable or very prominent, and the whole leaf surface is often covered with a greyish waxy 'bloom'. Some species have a violet suffusion on the base of the leaves and stem. There is also great variation in the texture, from thin and soft to tough, thick and rigid.

The stem

There is a considerable variation in the amount of stem development in the various species from more or less stemless ones such as *I. unguicularis* to short-stemmed ones like *I. pumila* and long-stemmed species such as *I germanica*. The stems may be simple (i.e. unbranched) or branched, and the branches may be very short, so that the lateral flowers are nearly sessile, to several centimetres long. Most species have rounded stems but in a few they are somewhat flattened or nearly two-winged. Some species have hollow stems, others solid ones.

The bracts

These are often also referred to as spathes or spathe valves. They are the membranous or papery organs enclosing and protecting the flowers in the bud stage. There are at least two, and sometimes more, enclosing or subtending one or more flowers. Their characters are worth noting for identification purposes, for their texture and colour can vary greatly from species to species and some have a sharp keel on the exterior. Some bracts are long and tubular, sheathing the perianth tube, while others are shorter and inflated or boat-shaped.

The flowers

The line drawings on p.3 show iris flowers of two widely different types with the various parts indicated. The terms shown are used throughout the book to describe the main features, such as the perianth tube, the ovary, falls,

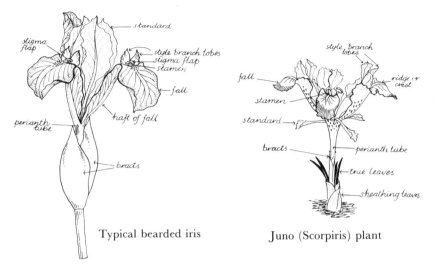

 components — labelled diagram with text:

Typical bearded iris:
standard
stigma flap
style branch lobes
stigma flap
stamen
fall
perianth tube
haft of fall
bracts

Juno (Scorpiris) plant:
style branch lobes
fall
ridge or crest
stamen
standard
bracts
perianth tube
true leaves
sheathing leaves

The parts of an *Iris* flower:

standards, style branches and stigma flaps. In some groups there is a beard in the centre of the falls while others have a cockscomb-like crest; some have only a smooth ridge. The colour varies enormously, either the whole flower being of uniform colouring or bicoloured; in the latter case the falls are normally darker than the standards. On the falls there is often a signal patch of a different or stronger colour which is presumably to attract pollinating insects into the pollination tunnel.

The capsules and seeds

These vary a great deal in shape and size and their characters are widely used in classifying the genus, especially in the Limniris subgenus. The capsules may be very short and dumpy to long and cylindrical, three-cornered in section or more or less rounded and they may have prominent or insignificant ribs running longitudinally. At the apex there is often a tapered portion, sometimes long, slender and beak-like and resembling a perianth tube.

The seeds are generally quite large and here too there is a great range of types. Some have a fleshy appendage known as an aril and this is important in grouping the species. The seed surface may be corky or smooth and sometimes with a loose papery and shiny coat.

These are the gross characters which determine the overall appearance of the plant. Obviously there are a great many others as well which could be, and often are, used in taxonomy; for example, the type of pollen and the chromosome number and structure. The scanning electron microscope is a new tool in the equipment of a botanist so that he can now study in much greater detail the outer coats of seeds and the architecture of the surface of pollen grains. In addition to these morphological characters the modern

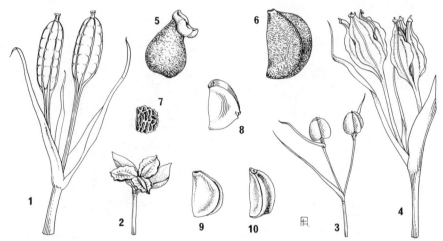

Capsules and seeds of *Iris*: 1 Capsules of *I. lactea*; 2 Capsules of *I. ruthenica*;
3 Capsules of *I. gracilipes*; 4 Capsules of *I. graminea*; 5 Seed of *I. korolkowii*; 6 Seed of
I. hexagona; 7 Seed of *I. bungei*; 8 Seed of *I. spuria*; 9 Seed of *I. versicolor*; 10 Seed of
I. versicolor

taxonomist should look at the whole aspect of a plant and its behaviour when
deciding its relationships or status. The habitat in which a plant has evolved is
highly important, as is its natural geographical distribution. The pollination
mechanism is undoubtedly also of great significance, for a species may well
rely on one particular pollinator for its very survival as a wild plant. Flower
colour is frequently scorned as a useful taxonomic feature, but to the pollinator
the colour or shade may be the main attraction!

Cultivation methods

I have commented about cultivation mainly in the introduction to each
Subgenus, Section or Series, or in the account of each species where there is
some special treatment worth noting. The species grow in a very diverse range
of wild habitats and it is difficult to generalize about their cultivation.
However, it is useful to look at the various ways in which irises can be utilized
in gardens, without particular reference to individual species. The comments are mostly based on my own experiences in southern England so
that readers in other parts of Britain will have to adapt them somewhat for
their own local conditions.

Obviously it is impossible to give precise cultivation details which apply to
all gardens in Britain and North America and for this reason particular
attention has been given to information concerning the natural conditions in
which irises occur in the wild. With this knowledge the gardener can
improvize in order to 'change' his own particular growing conditions to suit
the plants he is dealing with.

With the wide range of climatic conditions encountered in the United States it will be found that in some places certain groups of species are easy to grow whilst others are near-impossible. Thus, as an example, in the south-western dry-summer States such as southern California, New Mexico and Arizona the Oncocyclus and Scorpiris groups, which are often natives of semi-arid habitats, are comparatively easy to grow whereas in the much cooler northern States they are not very successful.

As a general guide to the use of this book by North American growers it is worth noting that the climatic conditions in the area where my own experience has been gained are roughly similar to those encountered in the north-west of the continent, especially the Vancouver region: that is, a fair amount of frost in winter, often coupled with a damp atmosphere, and relatively humid summers with a very variable light intensity. Most irises can be grown in these conditions but some need protection from the damp-cold in winter and others need covering in summer to keep off excess moisture while dormant.

Using this as a guideline it can be said that any region which has cold but dryish winters is acceptable for those irises which die down completely at that time of year, especially if a good snow cover can be relied upon. Conversely, regions of dryish summers with high light intensity are good for the semi-desert and mountain steppe plants of Asia, such as those of the Oncocyclus, Regelia, Scorpiris and Hermodactyloides groups which are dormant in summer and in fact require a warm dry period in order to flourish. Really ideal climates for growing the great majority of irises would combine cold, rather dry, winters with a moisture-rich spring and early summer followed by a warm dryish mid to late summer.

In general it is probably best to concentrate on those groups which 'do' really well in one's own particular area than to strive against the climate, although I find that the temptation to try 'the ungrowable' is too great—and I am sure that iris growers in America, Canada, Australia, New Zealand or anywhere else are no different in this respect!

Irises as border plants

This in general applies to the taller more robust species which can be planted among other herbaceous plants or shrubs and which do not require excessive moisture. Indeed, the taller bearded irises are better if the summer months are dryish so that their rhizomes become mature and well-ripened by the sun. Those of the Spuria and Sibirica groups on the other hand require more water in dry spells although they are ideally suited to mixed borders. Most irises need plenty of light and air and it follows that if they are planted in a mixed border then the accompanying species should not be placed too close to the irises since they will cast too much shade on them in the summer.

Irises can of course be grown together in beds of their own, and for the real enthusiast this is probably the best way since they can be given individual treatment. One must remember though, that the beds will be rather dull for most of the year.

For most of the stronger growing types, well-rotted compost or manure can be incorporated in the soil since this improves not only the general fertility but also makes it more friable and better-drained. Fertilizers may be given, and I regard it as a good policy to use a fairly slow-acting one such as bonemeal, or one of the proprietary granular fertilizers. Particular attention should be paid to potash and phosphate content rather than nitrogen, which encourages rapid soft growth.

Soils which are too acid can be improved by the addition of lime, preferably in the form of ground limestone or dolomite chalk. Badly drained and very heavy soils are not good for iris cultivation and need coarse grit and humus worked into them, and in really bad cases it may also be necessary to raise or slope the bed so that excess water drains away.

Irises coming into this category include the taller bearded ones; the two Lophiris (evansia) species *I. milesii* and *I. tectorum;* a few of the more robust Scorpiris (juno) species; the Sibirica group; the taller Spurias; *I. foetidissima; I. setosa; I. lactea* ('jensata') and the Spanish (Xiphium) group. If the soil is not alkaline some of the taller Pacific Coast irises (the Californicae) are also suitable. *I. unguicularis* is a satisfactory one for winter flowering but it must have the maximum amount of sun at all times and is best if planted in a protected site with a wall or fence behind it to trap warmth from the sun.

Irises for water and bog gardens

Some species are most decidedly water plants and no other position will do, whilst others, although naturally occurring in wet places, will tolerate border conditions if they are never allowed to become dry. A good humus content in the soil helps to retain moisture and if this is in the form of well-rotted manure or compost then it is sufficiently rich and no other form of feeding is necessary.

Irises which respond well to waterside treatment are mostly those of the Sibiricae, the Laevigatae and the Hexagonae Series although some of the species in the latter group are tender in Surrey. *I. prismatica* also grows and flowers better in moist, but not waterlogged, situations. Away from the waterside the Sibiricae are equally satisfactory, if the soil is humus-rich and moist in summer, as is *I. versicolor.*

Irises for rock gardens

Obviously here we are dealing with the smaller species which will look in place on a rock garden, although they are not necessarily mountain plants in the wild. The main requirements for most of them are again sun and drainage, the latter especially, since in cold wet winters many of the less robust irises succumb. A near-neutral or slightly alkaline soil is best with extra grit mixed in and not too much humus. There are however a few species which are dwarf and eminently suitable for rock gardens but which require a cool position in leafy or peaty soil; it is best to make up a separate section on the rock garden for these.

Those which require warm sunny positions include some of the species of the Hermodactyloides (reticulata) group, a few of the Scorpiris (juno) species—although most of these are bulb frame subjects—dwarf bearded species such as *I. pumila*, also *I. tenuifolia* and *I. setosa* in its more compact forms, and the smaller species of the Spuria group.

Those which need cooler damper conditions but still with plenty of light and good drainage (i.e. a peaty or leafy soil with grit added) include the small Lophiris (evansia) species *I. cristata*, *I. lacustris* and *I. gracilipes;* the unusual *I. kumaonensis;* the variants of *I. ruthenica;* and *I. verna.*

Several of the shorter members of the Pacific Coast group (Californicae) (e.g. *I. innominata*) are excellent rock garden plants if the soil is not alkaline. The advantage of a rock garden is that pockets of soil can be made with different soil mixtures to suit particular species.

Irises for peat gardens and light woodland

For the most part, the same comments apply to peat gardens as to rock gardens. My own peat garden consists of raised beds supported by dead tree trunks, or by peat blocks, and in these I grow plants which prefer cool growing conditions and slightly acid soils. I have found that several of the Pacific Coast irises and *I. ruthenica*, *I. prismatica*, *I. verna*, *I. cristata*, *I. gracilipes*, *I. setosa* and *I. foetidissima* are all ideally suited to this part of the garden, and to my surprise so is *I. japonica* which spreads and flowers freely. The Spuria, *I. sintenisii*, has also done well, planted near the roots of a ceanothus bush where it does not get too damp. The peat garden is top-dressed with leafmould or moss peat each winter at 'cleaning-up' time but does not receive fertilizers since growth seems to be quite vigorous enough without.

In lightly shaded areas or open woodland the same species can be grown and one can add the stronger growing Sibirica irises to the list as they are very suitable subjects for planting between, for example, rhododendrons, to provide some colour after the shrubs have finished flowering.

Irises for bulb frames

For enthusiasts wishing to grow the more specialized species of the Xiphium, Hermodactyloides (reticulata), Scorpiris (juno), Oncocyclus and Regelia groups it will be necessary to build some form of frame where the conditions can be controlled to a greater extent than in the open ground. It is not so much temperature but rainfall which needs to be controlled, for most of the species concerned are hardy in Britain. In their natural environment (almost all are from western and central Asia) they are subjected to long, hot, dry summers and cold winters, with a rather short growing season in spring when plenty of moisture is available. It is clear that, in Britain at least, we cannot be certain of a dry summer resting period to ripen the bulbs or rhizomes, or a cold enough winter to bring growth to a standstill. Thus we may have a cool wet summer when the plants tend to rot off, or there may be a long sunless wet autumn and winter when the plants are induced to keep growing and are soft and susceptible to sharp frosts. Therefore if we can provide a raised bed with well-drained soil which can be covered by glass frames when necessary there is instantly available a good measure of control of the climate. The frame lights are normally kept on throughout the year except in the spring when they can be removed through the main growing and flowering period. In the winter they keep the plants relatively dry and protected from the worst of the weather, while in the summer they allow the bulbs or rhizomes to be sun-baked and properly ripened. At all times a good free flow of air through the sides of the frame is necessary.

Exactly how the frame is built is not important. It can be of brick or breeze-block construction, or even made of old railway sleepers which is the case in my own garden. The frame lights can be of plastic or glass, in the form of flat 'Dutch lights' or of the more expensive type with metal frames such as 'Pluie' or 'Access' frames, which have the extra advantage of fairly high sides with sliding glass panels for extra ventilation. These panels also enable work on the plants to be carried out without removal of the frame, whereas Dutch lights or corrugated plastic sheets have to be removed each time access is required. It almost goes without saying that the bulb frame must be situated in as open and sunny a place as possible.

The soil mixture should be an open gritty one with good drainage but not too light and sandy, since this will dry out too much and contain very little nourishment. Excess humus must also be avoided, and it is best for these types of iris if the soil is slightly alkaline, brought about by the addition of dolomite chalk or powdered limestone rather than garden lime or ordinary chalk. After planting, a topping of about 1–2cm of coarse sand or chippings is of benefit since it discourages weeds and liverwort, and prevents the soil surface from becoming compacted at watering time.

Feeding is necessary after the first year and I prefer to scatter pellets of a granular general fertilizer such as 'National Growmore' on the surface of the soil in the autumn and spring. This dissolves slowly with each watering and the nutrients wash down to the roots.

The watering programme is simple, and is really a matter of providing plenty of moisture during the growing season, from about February to June depending upon the earliness of the spring. After this, water is withheld while the plants are dormant and then one good soaking is given in September or October while the temperature is still reasonably high — this stimulates the roots into growth, but not much leaf growth. No more watering is carried out through the winter until the plants begin their active spring growth. I think that the 'autumn drench' plays a useful role in getting the roots started into action. If it is left until later when the temperature has fallen the soil is often too cold for the roots to become active, even when supplied with moisture, and the plants stay quite dormant in cold wet soil for several months.

The labelling of a large specialist collection is essential and this is best done in conjunction with a planting plan of the frame; if any labels are lost then the plants can still be identified from the map.

Weeding is a problem and in the growing season it must be done by hand regularly or the weeds begin to shade the plants and prevent a free flow of air around them. In the dormant period they are less troublesome since the frame is dried off anyway and weed seeds are discouraged from germinating.

It is important to collect any seeds from the irises, not only so that the plants can be propagated but also to prevent self-sown seedlings appearing and causing a mix-up in the frame.

The bulb frame is an extremely satisfactory way of growing the more tricky irises and can be a worthwhile garden feature. Even with some of the common species, such as *I. danfordiae*, it is worth planting a patch in such a frame for they will often flower and increase much better than in the open ground.

Irises in pots

Undoubtedly pot cultivation gives the grower the maximum amount of control over the environment of any particular plant, although at the same time I am sure that many species are much better if they can be planted out in the open ground or in a bulb frame. Pots restrict the root activity a great deal and care must be taken over watering. However if a deep pot is used and it is buried up to the rim in sand then large fluctuations in the moisture content, and in the temperature, of the soil are reduced to a minimum. The pots I always use are clay pots since I have had little success with plastic ones. Roughly the same rules apply to pot cultivation as to bulb frames; the plants are given as much light and air as possible throughout the year — an alpine house or cold frame is ideal for this purpose — and the pots are dried out at the end of the growing period when flowering and fruiting is complete. Then after a warm dry summer dormancy they are repotted (every year) and watered in September to encourage root growth. A little moisture is given through the winter to prevent drying out, and then from about February onwards the plants are given as much water as they require. Being confined to a small volume of soil means that feeding is necessary. On the whole I find that a John Innes-type compost is satisactory, provided that it is well supplied with coarse gritty sand for good drainage. I prefer to feed with granules of National Growmore fertilizer, or something similar, scattered on the surface of the soil so that the nutrients become available over a period as they are washed down during watering. Liquid feeds are also suitable, avoiding those which are rich in nitrogen.

If the pots are plunged into sand there is no great danger during severe frosts, but I would not recommend leaving them in a cold greenhouse or frame if the pots are exposed all round to the air. Even hardy species can have their bulbs or rhizomes damaged if the soil is frozen right through.

Bulbous species are best potted with the top of the bulb covered by at least 3–4cm of soil, while rhizomatous ones have the rhizome at soil level. It does no harm, and provides some protection in winter, if the rhizomes are covered with coarse sand or grit and it does help to prevent the soil being washed out or compacted at watering time.

Any of the smaller species can be grown in this way and it is a particularly good method of cultivation for the really dwarf bearded (pogon) species, and the Oncocyclus, Regelia and Scorpiris (juno) groups. The 'reticulata' irises are very attractive as early spring alpine house plants but are on the whole more vigorous and persistent if grown in a bulb frame or rock garden.

Tender Iris species in greenhouses

Some of the irises are not hardy, even in southern England, and to grow these it is necessary to provide greenhouse or conservatory protection, where frosts can be excluded. In the main this applies to some of the more vigorous species of the Lophiris (evansia) group such as *I. wattii, I. formosana* and *I. confusa,* and to *I. decora (I. nepalensis)* and *I. speculatrix. I. confusa* is, I find, hardy in Surrey but always looks much better if given protection and I normally keep some plants in a tub which is stood on the terrace at flowering time in spring. The others are best planted into the greenhouse border where they have sufficient room to develop properly. *I. speculatrix* is small enough for a pot, as is *I. decora.* The composts are the same as those recommended above, but the point to remember about these irises is that they require watering throughout the year, with no real dormant period. They are also fairly vigorous and require feeding several times a year, either with a granular fertilizer as mentioned before, or with liquid feeds.

Propagation

The propagation of irises is by two means, either by division of the rhizomes or bulbs when the clumps have become large enough, or by raising new plants from seed.

Division is simply a matter of pulling the clumps apart and planting the resulting offspring. With some species this is a very simple matter and success can be almost guaranteed, whilst in others the divisions do not beome re-established very readily or quickly. The timing of the operation is fairly important and differs widely so I have commented upon this in the introductory remarks to the descriptions of each of the separate groups. In general it is best to divide them when new root activity is about to start and it is often possible to see this by scraping the soil carefully away from the rhizomes.

The advantage of vegetative propagation by division is that new flowering-sized plants are obtained quickly and they are all identical to the parent plant. Thus if a particular form is required it can be increased without the risk of variation.

Seeds on the other hand are likely to give rise to variable offspring, which can be advantageous if new forms are sought. Some species are slow to grow into clumps and therefore vegetative propagation is correspondingly slow, so that seeds in this case may present a more rapid way of increasing the stock. Another benefit of growing new young plants from seed is that they are free from virus whereas any infected plants which are divided will only give rise to virused offspring. If one possesses several different plants of a species it is almost certainly beneficial to cross-pollinate them to obtain seeds, although this must be done carefully to avoid hybridization with other species. My own aim is to perpetuate and increase if possible the true species rather than to hybridize them to produce new cultivars, although hybridization can result in some very fine garden plants. A lot of time and money often goes into the search for wild plants and I feel that crossing experiments should come much later, after a species has become well established and fairly widespread in cultivation.

Having obtained seeds, I normally sow them straightaway since this is the practice in nature. If they have to be kept for a period of time, I soak them for two or three days before sowing, changing the water several times during this period. They are best sown thinly in pots filled with a gritty soil mixture and the seeds are then covered with coarse grit which prevents growth of mosses and liverworts, and breaks up heavy rain which might wash them out. With nearly all the species I leave the pots plunged in a sand frame but exposed to the elements, since many of them seem to germinate in autumn or winter with the falling temperatures. Once germination has taken place they are given the protection of a cold frame or greenhouse for the rest of the first growing season. Subsequently they can be planted out into a bulb frame or the open ground, depending upon the species. Seeds of tender ones should of course be sown in a greenhouse.

In addition to the two traditional methods of propagation, there are some more modern techniques which involve the use of laboratory equipment and are probably beyond the scope of the average amateur. These are embryo culture and meristem culture.

The first, in which the embryo is removed from the seed and grown in an agar jelly culture, is important in that in skilled hands it can result in a high success rate from seeds of species which are normally difficult to germinate by conventional means — such as those of the Oncocyclus irises. It is considered beyond the purpose of this book to go further into this, but the interested reader can find much useful information in the papers by L. F. and F. Randolph (1955), Lee W. Lenz (1955) and P. Werckmeister (1955) (see Bibliography, p. 191).

The second 'special technique' in propagation is meristem culture whereby the minute growing point at the apex of the bulb or rhizome is removed and then cultured in laboratory conditions using plant growth substances. This method has been used to great advantage to grow virus-free offspring from an infected parent. Viruses are usually not present in the meristematic tip of plants and if this tissue can be removed and grown in isolation from the parent, a clean stock can be produced. Further information about this technique can be found in a paper by T. Murashige (1974).

Pests and diseases

I will not dwell on this subject for it is better on the whole to concentrate on growing the plants well rather than trying to cure sickly plants which have succumbed to diseases. Healthy strong-growing plants are the key to avoiding diseases. However, inevitably, the plants will occasionally get some trouble and here it is better to seek the advice of a plant pathology department, such as is operated by the Ministry of Agriculture, the Royal Horticultural Society or by Kew Gardens. Once the disease or pest is identified an attempt can be made to eradicate it. There are various leaf spot and rhizome rot diseases, ink disease of 'reticulata' group bulbs, and viruses. The first two are fairly readily dealt with by modern systemic fungicides in my experience, and by keeping plants cleared of dying leaves. Ink disease, which kills the bulbs, can be prevented by dipping the bulbs in a 'Benlate' solution at replanting time each year. Virus diseases are by far the most crippling for they cause the plants to lose vigour and look unsightly. They are spread by aphids so that regular treatment is necessary by spraying with an insecticide, preferably a systemic one, or by the use of fumigant smokes in a greenhouse. Plants already infected by virus are almost impossible to cure, although it can be done by meristem culture as described above under propagation techniques. Further useful notes on diseases and pests can be found in *Diseases of Bulbs* (Ministry of Agriculture Bulletin no. 117 by W. R. Moore), *Horticultural Pests* by Fox Wilson, revised by Dr. P. Becker, and in the Royal Horticultural Society's *Dictionary of Gardening*.

Common pests such as slugs are very damaging but are easily eradicated.

Classification

It is fortunate that the genus Iris with its many species can be divided into groups, making identification somewhat simpler. The methods of grouping vary somewhat and it is probably true to say that there are nearly as many different classifications as there are botanists who have studied the subject. However, most systems are basically the same—that is, each group or unit of species is fairly constant; it is just the status of each group which varies from system to system. Thus, for example, the 'juno' irises are classed as a separate genus, *Juno*, by Rodionenko; as a subgenus of *Iris*, known as Scorpiris by Spach; and as Section Juno by Dykes—all three of these have much the same content of species, although of course many new ones have been added over the years so that Rodionenko's group contains more than that of Dykes.

The grouping I have adopted in this book is based on the classifications of G. H. M. Lawrence (1953) and G. I. Rodionenko (1961). The most important way in which I have diverged from these is in the treatment of the bearded species, and for these I have followed the work of John J. Taylor (1976) who takes a new and to me very logical approach to the problems of classifying the subgenus Iris.

It is clear, from the recent work of P. Goldblatt (1980) that *Iris sisyrinchium* should be removed from the genus and reside in *Gynandriris*, while Lee W. Lenz (1972) shows convincingly that *Iris dichotoma* belongs to a distinct genus, *Pardanthopsis*

Key to temperate northern hemisphere genera related to *Iris*

1 Perianth consisting of six equal segments; style branches slender, not flattened

Belamcanda (page 186)

Perianth consisting of two whorls, the three outer segments with a spreading or reflexed blade ('falls'), the three inner segments ('standards') often smaller and erect, spreading or reflexed; style branches flattened and with a two-lobed apex   2

---

2 Rootstock consisting of finger-like tubers; leaves square in cross-section; flowers green and blackish; ovary unilocular   *Hermodactylus* (page 183)

Rootstock various but not as above; leaves various but if square-sectioned, then associated with a net-coated bulb; ovary three-locular   3

---

3 Rootstock a small corm with a netted tunic; leaves channelled; flowers with no perianth tube
Gynandriris (page 181)

Rootstock various but if with a netted tunic, then bulbous and associated with square-sectioned leaves; perianth tube usually distinct but if not, then plant has flat leaves   4

---

4 Perianth tube absent; lower part of the falls (haft) marked with transverse bands; each flower lasting only one day   *Pardanthopsis* (page 184)

Perianth tube present, although sometimes very short; haft of falls if marked at all then with longitudinal lines or spots; flowers usually lasting for several days   *Iris* (page 19)

A classification of *Iris*

1 Iris subgenus Iris
   A   Section Iris (The bearded or pogon irises) (page 19)
   B   Section Psammiris (Spach) J. Taylor (page 38)
   C   Section Oncocyclus (Siemssen) Baker (page 40)
   D   Section Regelia Lynch (page 61)
   E   Section Hexapogon (Bunge) Baker (page 64)
   F   Section Pseudoregelia Dykes (page 66)
2 Iris subgenus Limniris (Tausch) Spach (The beardless irises)
   A   Section Lophiris (Tausch) Tausch (The evansia irises) (page 69)
   B   Section Limniris
      (a)   Series Chinenses (Diels) Lawrence (page 78)
      (b)   Series Vernae (Diels) Lawrence (page 81)
      (c)   Series Ruthenicae (Diels) Lawrence (page 82)
      (d)   Series Tripetalae (Diels) Lawrence (page 84)
      (e)   Series Sibiricae (Diels) Lawrence (page 87)
      (f)   Series Californicae (Diels) Lawrence (page 92)
      (g)   Series Longipetalae (Diels) Lawrence (page 101)
      (h)   Series Laevigatae (Diels) Lawrence (page 103)
      (i)   Series Hexagonae (Diels) Lawrence (page 105)
      (j)   Series Prismaticae (Diels) Lawrence (page 108)
      (k)   Series Spuriae (Diels) Lawrence (page 109)
      (l)   Series Foetidissimae (Diels) Mathew* (page 120)
      (m)   Series Tenuifoliae (Diels) Lawrence (page 121)
      (n)   Series Ensatae (Diels) Lawrence (page 125)
      (o)   Series Syriacae (Diels) Lawrence (page 127)
      (p)   Series Unguiculares (Diels) Lawrence (page 129)
3 Iris subgenus Nepalensis (Dykes) Lawrence (page 132)
4 Iris subgenus Xiphium (Miller) Spach (page 134)
5 Iris subgenus Scorpiris Spach (The juno irises) (page 138)
6 Iris subgenus Hermodactyloides Spach (The reticulata irises) (page 171)

Key to subgenera of *Iris*

1 Plants with a bulb; leaves channelled, quadrangular or almost cylindrical   2
Plants not bulbous, usually with a slender to fat rhizome; leaves flat but may be very narrow   4

---

* Iris subgenus Limniris Section Limniris Series Foetidissimae (Diels) Mathew [Iris subsection Foetidissima Diels in Engl. & Prantl, *Nat. Pflanzenfam. Aufl.* **2**, 15a: 502 (1930)].

2 Bulb tunics fibrous and netted; leaves four-sided or nearly cylindrical in section (except in *I. kolpakowskiana*)   Subgenus Hermodactyloides (page 171)
Bulb tunics papery to tough and leathery, never netted-fibrous; leaves channelled   3

3 Bulbs usually with thickened fleshy roots present when dormant; standards much-reduced in size (except in *I. cycloglossa*)   Subgenus Scorpiris (page 138)
Bulbs with thin fibrous roots, usually dying away at dormancy; standards not greatly reduced (except in *I. serotina*)   Subgenus Xiphium (page 134)

4 Plant with tiny rhizomes and markedly swollen tuber-like storage roots as well as fibrous ones   Subgenus Nepalensis (page 132)
Plant with well-developed rhizomes, although sometimes very slender and stolon-like; roots not swollen and tuber-like   5

5 Falls with a distinct beard of fairly long hairs   Subgenus Iris (for key to Sections, see page 15)
Falls without an obvious beard but may have cockscomb-like crests, or rarely a fine pubescence of very short unicellular papillae on the haft   Subgenus Limniris (for guide to Section and Series, see page 15)

Key to sections of subgenus Iris

1 Seeds with a fleshy appendage (aril); stems never branched; hairs on falls unicellular   2
Seeds with no fleshy appendage; stems often branched; hairs on falls multicellular   Section Iris (page 19)

2 Flowers in shades of lilac, purple or blue, conspicuously mottled and blotched darker   Section Pseudoregelia (page 66)
Flowers of various colours, sometimes veined or finely spotted but not mottled or blotched   3

3 Falls and standards provided with a beard   4
Falls only bearded   5

4 Each stem normally with two flowers produced from two bracts; mountain plants   Section Regelia (page 61)
Each stem with three or more flowers produced from three or four bracts; semi-desert plants   Section Hexapogon (page 64)

5 Flowers only 3–5cm in diameter; stems with one to three flowers; flowers yellow (sometimes lavender-purple in *I. potaninii*)   Section Psammiris (page 38)
Flowers at least 5cm in diameter, usually much more; stems one-flowered; flowers various colours but rarely yellow *(I. auranitica* and *I. barnumae f. urmiensis)*   Section Oncocyclus (page 40)

Guide to sections and series of subgenus Limniris

The following guide is not intended to be a formal key. I have purposely not made a 'proper' key which selects out a few characters since the classification of this subgenus is largely based on features of the capsules and seeds, and on stigma type—all rather difficult points to observe. The guide, on the other hand, takes into account distribution and habitat as well as other features in the hope that this will assist in identification.

It is recommended that the user of the guide tries each description in turn and chooses that which agrees most with the plant in question.

Section Lophiris ('evansia' irises, page 69)

Falls with one or more ridges or crests, usually dissected like a cockscomb (not in *I. tenuis* from Oregon or *I. speculatrix* from Hong Kong/China); leaves soft in texture; woodland plants often with widely-spreading rhizomes. East Asia, Oregon and east-central United States.

Section Limniris

1 Seeds scarlet, remaining attached to the capsule after it has split open; plants evergreen with wide leaves. West Europe and North Africa   Series Foetidissimae (page 120)

---

2 Rhizome nearly vertical, covered with spiny needle-like fibres. South Turkey, Syria, Jordan, Iraq and Israel   Series Syriacae (page 127)

---

3 Standards much-reduced, often to bristle-like proportions. East Asia, Alaska and Canada
Series Tripetalae (page 84)

---

4 Small woodland plant with blue flowers from south-east United States; capsule three-cornered, containing globose seeds which have a fleshy appendage when fresh
Series Vernae (page 81)

---

5 Plant of wet grassy habitats in east United States; stems tall with brownish bracts; capsule three-cornered; seeds smooth; rhizome thin and widely-creeping Series Prismaticae (page 108)

---

6 Low plants, but with perianth tubes 6–16cm long; style branches with golden glands on the margins. Mediterranean and Black Sea region   Series Unguiculares (page 129)

---

7 Capsule rounded in section, only 1–1·5cm long, with the three valves curling back and quickly releasing the seeds which are pear-shaped and have a fleshy appendage; plant low and compact-clump forming; perianth tube about 1cm long. Asiatic.
Series Ruthenicae (page 82) & *I. speculatrix* (page 75)

---

8 Plants with thin wiry, often widely-creeping rhizomes; capsules triangular; leaves prominently ribbed. East Asia.   Series Chinenses (page 78)

---

9 Apex of rhizome with shiny brown leaf bases; perianth tube 4–14cm long; stigma bilobed; capsules cylindrical or shortly ellipsoid, often with six equally spaced ribs; seeds wrinkled on surface. Asiatic steppe plants   Series Tenuifoliae (page 121)

---

10 Plants from margins of irrigation ditches and salt marshes; perianth tube only 2–3mm long; ovary six-grooved and beaked at the apex; capsule with six prominent ribs and shiny globose seeds. Central and east Asia   Series Ensatae (page 125)

---

11 Capsules with two ribs at each of the three corners; stigma two-toothed; seeds with a loose shiny coat. Europe and Asia   Series Spuriae (page 109)

---

12 Very robust plants from marshes and swamps; stems with leaf-like spathes and large flowers; capsules six-ribbed; seeds very large and corky. South United States

Series Hexagonae (page 105)

---

13 Robust moisture-loving plants; capsules three-cornered or nearly cylindrical in section with three ribs, thin-walled and breaking up irregularly rather than splitting into three; seeds shiny; stigma bilobed. Europe, Asia, east United States

Series Laevigatae (page 103)

---

14 Plants from damp places or alpine meadows; stems hollow (except *I. clarkei*); stigma triangular; capsules three-cornered or nearly round in section; seeds D-shaped; falls with two flanges at the base of the haft. Europe to east Asia   Series Sibiricae (page 87)

---

15 Plants from acid to neutral dryish places in the Pacific Coast States of N. America; stigma triangular (but squared or bilobed in *I. purdyi*); capsule with three ribs, three-cornered or cylindrical; leaves tough   Series Californicae (page 92)

---

16 Plants from calcareous soils which are very wet in winter and spring; stigma two-toothed; stems persisting for a year or more after flowering time; capsules six-ribbed, tapering at both ends. Central and west United States   Series Longipetalae (page 101)

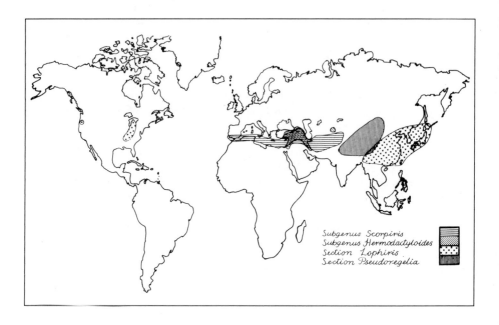

Subgenus *Scorpiris*
Subgenus *Hermodactyloides*
Section *Lophiris*
Section *Pseudoregelia*

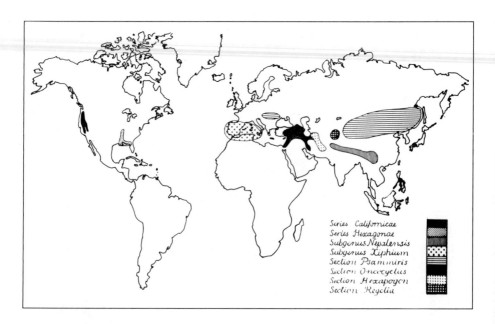

Series *Californicae*
Series *Hexagonae*
Subgenus *Nepalensis*
Subgenus *Xiphium*
Section *Psammiris*
Section *Oncocyclus*
Section *Hexapogon*
Section *Regelia*

Maps showing distributions of some subgenera, sections and series of the genus *Iris*

The species of Iris

1A   Subgenus Iris, Section Iris (The bearded or pogon irises)

These are the well known bearded species which have given rise to the mass of colourful cultivars we now have available for our gardens.

Their characteristic features are: stout rhizomes giving rise to fans of sword-shaped, usually rather broad, leaves; simple or branched stems with two to several flowers, although in a few species the inflorescence is reduced to a single short-stemmed or stemless flower, as in *I. pumila* and *I. attica*. The flowers have well developed falls and standards in all species and there is of course a prominent beard in the centre of each of the three falls. The seeds have no fleshy appendage on them which is the feature separating Section Iris from the other sections in the subgenus (Oncocyclus, Regelia, Psammiris and Hexapogon).

It appears that there are not a great many wild species of bearded Iris, but the true position is somewhat obscured by the presence of many hybrids, some of which are well established in semi-wild situations and give the impression of being 'good' species. The taller, more robust bearded irises are striking plants of great ornamental value and they are also very easy to propagate vegetatively so that a great many individuals can be raised from one original hybrid or selection. Such is probably the case with the common *I. germanica* which is now grown on an almost world-wide scale and is often found in waste places, although usually in or near cultivated land. It can persist for a very long time by vegetative means but on the whole the plant does not spread far without the help of man. The same almost certainly applies to other less well-known 'species' such as *I. biliottii* which is known only from graveyards in Turkey and *I.* 'Florentina', the white-flowered variant of *I. germanica* which has been grown in Italy for the production of Orris Root. This yields a perfume not unlike that of violets. In western Asia bearded irises are often planted in cemeteries—it is said that the sword-like leaves are intended to drive away evil spirits trying to enter the graves. The spread of *I. albicans*, a wild species from Yemen and Arabia, can probably be attributed to pilgrims returning from Mecca, so that now this plant is very widespread in the Moslem world, nearly always in and around graveyards.

Many of the early hybrids to be cultivated were undoubtedly crosses between *I. pallida* and *I. variegata*, both European species, and W. R. Dykes observed hybrid populations of these in the wild. Many specific names which are in existence can be excluded from a list of iris species since they have been attached to plants of hybrid origin; sometimes these hybrids are of only sporadic appearance. Since my aim here is to give information about the wild species, I have made only a passing reference to the majority of these named hybrids—sufficient only for the reader to be able to account for any particular latinized epithet. I am not therefore including descriptions of plants such as *I. sambucina* and *I. squalens* but there will be a reference to their identity. After all, these plants have little more claim to specific identity than any of the modern cultivars. *I. germanica* is a little more difficult to dispense with since it is such a widespread and well-known plant, and is the type 'species' of the Section Iris. It is quite impossible to say whether it is a wild species or not and it seems best to leave well alone and include it, but accepting that it may itself be a hybrid of some antiquity.

Most of the bearded iris species are in cultivation in Britain but the garden hybrids and selections are based on only a few of these. It is my hope that there will always be enough 'species enthusiasts' to maintain the true plants so that their identity is not lost. It is, for example very difficult to obtain an authentic *I. pumila*, most of those offered being 'dwarf beardeds' of hybrid origin. We must, in growing the species, bear in mind the conservation question, for with the destruction of their natural habitats, and the fact that they hybridize so readily, it would be all too easy to find some of the species in an endangered state of existence.

Cultivation

Much has been written on the cultivation of bearded irises and the reasons for this are easy to understand if one has tried to grow a collection of them; they are not as easy as one might think, setting aside a few of the taller species such as *I. germanica* and *I. pallida* which will thrive in almost any sunny situation.

It is possible to generalize and say that they need well-drained soils, preferably slightly alkaline and with a clay content rather than light sand, although if it is too heavy some grit should be added to improve the drainage. In acid soils some lime should be incorporated, preferably in the form of ground limestone or dolomite chalk. The more robust species will be improved by the addition of humus to the soil before planting, in the form of well-rotted leafmould, moss peat or very old manure. For many of the dwarf species raised beds are preferable for extra drainage, and rock gardens are ideal for providing such a position.

The situation should, for all species, be in full sun, for although some will tolerate shade they do not usually flower very well.

Planting should be carried out in August or September since most of the species start new root activity soon after this. With the larger types it is best to clip their leaves down to about 15-20cm long to reduce transpiration and minimize the risk of being blown over by strong winds before the new roots have anchored the rhizomes. The rhizomes must be planted so that their upper surface is exposed—burying them results in a lack of flower and probable loss due to rotting. In spring a feed of fertilizer can be given but the exact formula of this is not critical. I normally use a ready-mixed proprietary brand such as National Growmore which is granular and the nutrients are therefore released slowly over a period of time. Excessively rich nitrogen-based fertilizers should be avoided.

Some species are undoubtedly tender in Surrey, for example *I. attica*, and these are best grown as pot plants in an alpine house or cold frame, or in a raised bed covered with glass in winter.

Propagation is by division of established clumps or by seed. Old plants can be divided into single rhizomes with a fan of leaves. This is best done in August or September, usually every second or third year with the vigorous types. Even the smaller species often require frequent replanting since they appear to exhaust the soil and lose vigour quite rapidly.

Seeds of the bearded iris species germinate quite well and the young plants take two or three years to reach flowering size. The seedlings are best planted into prepared beds and left undisturbed until fully grown.

The species of subgenus *Iris* (the pogon or bearded iris)

I. aequiloba Ledeb. A synonym of *I. pumila*.

I. albertii Regel. This is a species I have not had the opportunity to grow yet, so my description is based on the observations of others, and on herbarium material. It is of medium (30-70cm) height when in flower with broad erect grey-green leaves which are tinged with purple at the base. The outer leaves at least are rather short and blunt-tipped. The stem has two or three branches and these carry one to three flowers, produced from slightly inflated bracts which are green with only slightly papery margins. Each flower is about 6-8cm in diameter and the colour varies in the wild from lavender to purplish-violet with brownish-red veining on the haft of the falls. The beard consists of whitish or pale blue hairs, tipped with yellow. Rodionenko describes a variant (forma *erythrocarpa*) in which the developing capsule is red.

I. albertii is an early flowering pogon iris, usually flowering in May, and is a native of Kazakstan in central Asiatic Russia where it grows on grassy steppes in the mountains at 1700-2000 metres altitude, particularly in the Tien Shan and Fergana ranges. It is a more slender-looking plant than *I. germanica* with smaller flowers. W. R. Dykes makes the observation that the veining on the haft of the falls ends abruptly in a straight line across the blade where the beard ends. I do not know if this is a consistent feature in the species.

I. albicans Lange. There has frequently been some confusion between this species, which occurs wild in Arabia, and 'Florentina', the albino version of *I. germanica*. In 1802 it was beautifully figured by Redouté under the name of *I. florentina* and it was not until 1860 that it was given specific status, when Lange named it *I. albicans*.

It is a stocky plant, usually 30-60cm in height in the wild, with short broad (1·5-2·5cm wide), very grey-green overwintering leaves which are abruptly narrowed to an incurving tip. The stem has one to three terminal flowers and is often unbranched but may have one sessile lateral flower head. The sweetly scented flowers are about 8-9cm in diameter in either pure white or blue, produced from very broad blunt bracts which are green or purplish tinted in the lower half or two-thirds, and papery-transparent in the upper part. The beard is of white hairs, tipped with yellow, and there is a greenish-yellow flush to the hafts of both the falls and standards. The blue-flowered form has been given the name 'Madonna'.

'Florentina' is easily distinguished from *I. albicans* because the bracts are almost wholly brown and papery at flowering time, the lateral flowers are stemmed and the flowers are very slightly flushed with pale blue, not as pure a white as in the white form of *I. albicans*. Additionally, 'Florentina' has narrower leaves which are greener in colour.

I. albicans is a native of Saudi Arabia and the Yemen Arab Republic where it grows in dry rocky places or on banks at altitudes up to 2700 metres. New material has been collected in recent years and introduced into cultivation and it seems to be a 'good' wild species. Both blue and white forms occur in the natural populations, although the latter seems to be much the most common. It seems fairly certain that *I. albicans* has been distributed through western Asia by man as a plant to adorn graveyards.

I. × alto-barbata E. Murray. This is a name originally intended to embrace the tall bearded garden irises, but not commonly used.

I. alexeenkoi Grossheim. There is very little difference between this and *I. pumila* and Rodionenko comments that it is distinguished only by having larger leaves (11-12cm long and 1·5cm broad) and larger flowers (7-8cm diameter). It is about 20-30cm in height and has solitary purple-blue flowers with a tube about 9-10cm long. The falls have a yellowish beard. It was described from eastern Transcaucasia in Russia where it grows in steppe country and flowers in April. The plants I have cultivated from time to time have all produced flowers of a fine deep colour. Unfortunately they have all rapidly succumbed to a virus disease which renders the plants very unsightly and of no further value.

I. amoena DC. The plant originally described by De Candolle and figured by Redouté had certain characteristics which suggest that it might have originated from *I. variegata* perhaps by selection or hybridization in gardens. Nowadays the name is applied to a horticultural group of iris cultivars which have white standards and coloured falls, not necessarily blue as in the original.

I. aphylla Linn. (Syn. *I. benacensis; I. bohemica; I. melzeri; I. nudicaulis*). The name of this species probably refers to the somewhat leafless-looking long branches of the inflorescence rather than to the fact that the plant is dormant without leaves in winter. Many bearded iris species have quite well-developed leaves in winter, so this is a worthwhile character to note. *I. aphylla* is a smallish species, about 15-30cm in height with curved outer leaves and erect inner ones, about 0·5-2cm wide. The flower stems are branched at or below the middle, often almost from the base, so it does have a rather different appearance from most of the species in the section in which any side branches arise well above the middle of the main axis. The flowers vary in number from one to five and they are pale to dark purple or violet-blue and about 6-7cm in diameter. The somewhat inflated bracts are 3-6cm long and are green, often with purplish tips.

I. aphylla is a central and eastern European plant recorded from the French Alps (see *I. perrieri*), eastwards through Hungary and southern Poland to western Russia, and to the north Caucasus if one includes with it *I. furcata*. *I. furcata* is usually treated as a synonym but the Russian iris authority Dr G. I. Rodionenko regards it as distinct, although the differences are not very great. It is said to have stems which branch somewhat higher up than in *I. aphylla*, not near the base, and it consistently has smaller seeds. In the *Iris Year Book* of 1967, Rodionenko writes 'Its flowers are smaller but of a better form, the colour deeper, the stem thinner and more elegant as is also the foliage than in *I. aphylla*' (translated by M. Rogoyski). The distribution of *I. furcata* is given as the northern Caucasus, Moldavia and southwestern Ukraine. *Flora Europaea* regards it as a straight synonym of *I. aphylla*, as did W. R. Dykes.

I. aphylla is a pleasing and distinctive little iris but I have found it extremely susceptible to virus. It readily produces seeds, however, and new clean stock can be raised without difficulty.

I. astrachanica Rodionenko. This name applies to a group of variable dwarf irises occurring in the area between the river Volga and the Ural mountains. Rodionenko makes the suggestion that they may be hybrids between *I. pumila* and *I. scariosa*.

I. attica Boissier & Heldreich. This delightful miniature iris is to me one of the most tantalizing, for it will sometimes grow vigorously for several years then collapse and disappear in a short space of time. In much of Britain and in the colder, damper States of North America, it is probably too cold and damp in winter for it to be reliably hardy and it is better to take the precaution of keeping a pot of each of the various colour forms in a frame through the worst of the months.

It is a very dwarf plant, more or less stemless and only 5-10cm in height when in flower. The strongly falcate leaves are only about 4-7cm long and 4-7mm wide. Otherwise it is fairly similar to the more well-known *I. pumila*, although smaller, with a flower diameter of about 3·5-4·5cm. As in this species there is a great range of colour variation in yellows or purples and there are bicoloured forms. The two bracts subtending the solitary flowers are rounded on the backs and they sheath the long perianth tube (5-7cm); occasionally the outer bract is somewhat keeled, but never both of them as in *I. suaveolens (I. mellita)*.

I. attica occurs wild in rocky places on limestone formations in Greece, where I have seen it in Attica and the Peloponnese, southern Yugoslavia, and locally in western Turkey, flowering in April or May. I remember with great pleasure a sunny mountainside in Attica some years ago when I saw *I. attica* for the first time in flower. There were hosts of forms from small purple-flowered ones with bright-blue beards on the falls to ones with much larger purple flowers; there were pale yellows to deep mustard, bicoloured ones with yellow standards and brownish or bluish falls, and one delightful variant which had wholly creamy-white flowers except for a blue blade to the falls. Detached pieces I brought home grew well and I still have some of the forms after a period of 15 years. Unlike several of the dwarf bearded species *I. attica* does not, in my experience, readily succumb to virus diseases.

I. australis Todaro. A synonym of *I. germanica*.

I. babadagica Rzazade & Goln. This dwarf species is little known as yet and is considered by Rodionenko to be related to *I. furcata*. It is about 10–15cm in height and has two purple-violet flowers per stem. The distribution is given as Babadag, a mountain in the eastern Caucasus.

I. balkana Janka. This is probably one of the variants of *I. reichenbachii*.

I. × barthii Prodan. A hybrid between *I. pumila* and *I. aphylla*.

I. × barthiiformis Prodan. A hybrid between *I. pumila* and *I. aphylla*.

I. bartonii M. Foster. A synonym of *I. kashmiriana*.

I. belouinii Boissier & Cornault. See under *I. germanica*.

I. benacensis A. Kerner. A synonym of *I. aphylla*.

I. biflora Linn. This is a confused name covering both *I. pumila* and *I. aphylla.*

I. biliottii M. Foster. See under *I. germanica*.

I. × binata Schur. Hybrids between *I. aphylla* and *I. pumila*.

I. bohemica P. W. Schmidt. A synonym of *I. aphylla*.

I. × borzae (Prodan) Prodan & Nyárády. A hybrid between *I. variegata* and *I. albicans*.

I. bosniaca G. Beck. A synonym of *I. reichenbachii*.

I. × cavarnae Prodan. A hybrid between *I. variegata* and *I. pallida*.

I. cengialtii Ambr. A synonym of *I. pallida* subsp. *cengialtii*.

I. chamaeiris Bertol. A synonym of *I. lutescens*.

I. cretica Hort. (Not *I. cretica* Herbert, which is a synonym of *I. cretensis*, an Unguiculares Series iris). This name was given to a dwarf rosy-purplish flowered pogon iris, probably an *I. pumila* variant, which was thought to have come from Crete. It has been successfully used in hybridization to add to the range of miniature iris cultivars. I can find no authentic records of *I. pumila* or *I. attica* from Crete.

I. croatica I. & M. Horvat. A synonym of *I. germanica*.

I. cypriana Baker & Foster. See under *I. germanica*.

I. dacica Beldie. A synonym of *I. aphylla*.

I. daesitiatensis. A name which has been used in literature but without details of the plant or author. It is not validly published under the Rules of Nomenclature as far as I can ascertain, and should be ignored.

I. dalmatica Hort. A synonym of *I. pallida*.

I. dragalz Horvat. A synonym of *I. variegata*.

I. erratica Tod. A synonym of *I. lutescens*.

I. eulefeldii Regel. A synonym of *I. scariosa*.

I. flavescens Delile. The plate of this by Redouté (t. 375) in *Les Liliacées* suggests that it is just a form of *I. variegata* in which the purple veining on the falls is rather slight.

I. florentina Linn. It seems to me that this is not a true wild species, but an albino cultivar, very closely linked with *I. germanica*. Linnaeus, in his original description of 1762 said that it was similar to *I. germanica* but with white flowers, although at the same time he referred to Miller's figure number 154 which represents the Spuria Iris, *I. orientalis* (= *I. ochroleuca*). It is clear however, that Linnaeus intended the name 'Florentina' for the well-known bearded florentine iris used in the perfume industry, particularly in Italy. I can see little justification for regarding this plant as anything more than a horticultural selection, in which case the name *I. germanica* 'Florentina' would be appropriate (see below).

The other commonly cultivated white-flowered pogon iris is *I. albicans*. This, I believe, is a true wild species in Arabia and Yemen, spread by man through the Moslem world. It is easily distinguished from 'Florentina' by several characters which are mentioned under the description of *I. albicans* above.

I. furcata M. Bieb. See comments under *I. aphylla*.

I. germanica Linn. (Syn. *I. croatica*). The 'German iris' is certainly by far the best known of the tall bearded irises, frequently cultivated and much more persistent than most of the species or cultivars. In favourable climates it becomes naturalized and is recorded from most parts of the world so that its true distribution (if indeed it occurs anywhere as a wild plant) is unknown. Almost certainly it is either a native of the Mediterranean region or is a hybrid of some species which occur there. Whatever its status it can be said that it is an ancient, very tenacious and useful garden plant capable of surviving a degree of neglect which dispenses with most of the tall bearded cultivars. Although it seldom produces seeds in gardens, I imagine this is due to unfavourable weather conditions at pollination or fertilization time rather than true sterility for it does form capsules and seeds in Mediterranean districts.

I. germanica is somewhat variable in its flower colour but is otherwise fairly uniform in its characters. It grows 60–120cm in height, the stem with one or two branches which are at least 5cm long. The grey-green leaves are 2·5–4·5cm wide and 30–40cm long. Each flower is about 9–10cm in diameter and is a variable shade of bluish-violet with a yellow beard. A very commonly seen form has the standards paler than the falls which are of a slightly more purplish shade. The bracts are rather short and broad, green or purplish in the lower half and a semi-transparent brownish colour in the upper half.

It is not easy to distinguish *I. germanica* from several other similar plants from the Near and Middle East which have been described as species. These are doubtfully wild and are recorded mostly from cemeteries or near habitation. The following notes are intended only as a guide to their identities and do not necessarily imply that I consider them to be distinct! The majority are most likely to be cultivars which have been propagated vegetatively and distributed locally, giving the impression of being stable populations with a distinct geographical range.

I. belouinii Boissier & Cornault. I am not familiar with this plant and can only pass on information from literature. It is apparently similar in general appearance to *I. germanica* but is reputed to have leaves which go dormant in winter. It is recorded from Morocco, between Fès and Meknès.

I. biliottii Foster. This grows in the Black Sea region of Turkey, especially in the valley leading from Trabzon to Gümüşhane where it is much used on graves. It has stems about 60–80cm in height with two or three branches 5–10cm long. The acute bracts are grey-green with transparent papery tips. Like *I. germanica* the flowers are scented and purple, the falls a rather reddish-purple and the standards of a bluer-purple shade. The haft of the falls is veined brown-purple on a white ground and the beard is white with yellow tips to the hairs.

I was shown this plant growing near Gümüşhane by T. Baytop and can confirm that it is apparently confined to cemeteries and does not produce seeds. Rhizomes collected by us were brought back to England where they grow vigorously and flower freely a few weeks after *I. germanica*.

I. cypriana Baker & Foster. This is again very similar in its general appearance and flower colour to *I. germanica* but can be distinguished by the outer bract being brown and wholly dry and papery. The flowers in the plants I have seen are enormous, about 15cm in diameter, nearly unscented and with markedly wedge-shaped falls, broadest at the apex as mentioned by Dykes. It is known from several localities in Cyprus, growing at the edge of fields, in waste places and in gardens, especially in the Troodos region. There is also a possibility that it is represented in southern Turkey, but from dried specimens one cannot be certain.

I. junonia Schott & Kotschy. This seems to be one of the most likely candidates when considering which of these 'germanica-like' plants are truly wild. It is about 50–65cm in height with about four branches to the inflorescence, the lowest of which is up to 10cm long. The grey-green leaves are much shorter than the flower stems, usually only about 30–35cm, and are 4–5cm wide, the widest point being at about the middle. The bracts are pale green in the lower half and transparent in the upper half. Although the flowers are similar in shape to those of *I. germanica*, they are smaller and very variable in colour from white to pale cream and deep yellow, and from pale blue to purple. The beard is yellow and the haft often veined brown-purple on a white ground.

I. junonia was first collected in the Cilician Taurus mountains and has been recollected on a few occasions. Dykes notes that its leaves die away in winter whereas those of *I. germanica*, and the others mentioned here, begin to grow in the autumn and are visible during the winter months.

I. mesopotamica Dykes (Syn. *I. ricardii*). This has flowers similar in overall colour to the others mentioned above and is very like *I. cypriana* but has bracts which are dry and papery only in the upper third, the lower part being thin-textured but green. Its flowers have a shorter perianth tube (1·3–1·8cm) than *I. cypriana* (2·5cm). It is a vigorous plant with long-branched stems 90–120cm in height and deep green, only slightly greyish, leaves 5cm wide. It has been recorded in several localities from southern Turkey southwards through Syria to Israel.

I. trojana Kerner ex Stapf. Although much the same in general appearance as all the rest of this group, *I. trojana* has a few features which make it a distinct and very fine garden plant. It is about 60–70cm in height with two or three long branches on the stem and rather narrow bracts which have a purplish-flushed base and brown papery tips. The sweetly scented flowers have reddish-purple falls with white, yellow-tipped beard hairs and paler, bluer standards. W. R. Dykes notes that the veining on the haft of the falls consists of much finer brown-purple lines than in the others and I find that this is certainly the case in the plants given to me by T. Baytop, collected in western Turkey. Dykes also regards the purple-stained long narrow buds as distinctive.

In addition to the above 'species' there are several named variants of *I. germanica* which one might encounter in literature and/or gardens.

I. germanica 'Amas' (Syn. var. *macrantha*) 'Amas' is a sturdy variant with deep blue-purple falls and rounder paler blue standards. There is a very prominent beard of bluish-white hairs tipped with orange. Dykes writes that the leaves die away and do not start to develop until spring. It was introduced to cultivation from the town of Amasya in northern Turkey.

I. germanica 'Askabadensis'. This variant came from Russian Central Asia and is not, as far as I can tell, now in cultivation in Britain. From herbarium material it appears to be a tall form and is said to be the latest of the 'germanicas' to flower. The falls are reddish-purple and the standards are a lighter pale blue. On the haft of the falls there is some yellowish-brown veining, which is not very conspicuous, and a yellow-tipped beard.

I. germanica 'Florentina' (Syn. *I. florentina* L.) This is the white-flowered variant of *I. germanica* which is frequently confused with *I. albicans*. 'Florentina' has beautiful, scented, flowers of a very faintly bluish shade rather than pure white and on the falls there is some greenish-yellow veining on the haft, and a deep yellow beard. Like *I. germanica* the inflorescence is branched and these branches are quite long, immediately distinguishing it from *I. albicans* in which the lateral flowers are sessile.

I. germanica 'Kharput'. This is named after the village of Harput in eastern-central Turkey from where it was originally introduced to cultivation. Its most interesting features are the reddish-purple edges to the leaves and the very dark almost blackish-purple, narrow falls. The standards are paler reddish-purple.

I. germanica 'Nepalensis' (Syn. *I. nepalensis* Wallich; 'Purple King'; 'Atropurpurea'). Unlike most of the *I. germanica* forms mentioned, 'Nepalensis' has the falls and standards almost the same shade of deep red-purple. The beard hairs are bluish-white at the apex of the beard, graduating to white with orange tips on the haft. It was brought into western gardens from Katmandu in Nepal.

I. germanica 'Sivas'. This was introduced from northern Turkey and is said to have bluer flowers than most *I. germanica* forms, with a blue ground colour to the haft so that the veining is less conspicuous than in those with a white ground. The beard is whitish with very little yellow colour on the hairs. I have not seen living material.

I. glaucescens Bunge. A synonym of *I. scariosa.*

I. glockiana O. Schwarz. This was described from material collected in the Izmir area of western Turkey. It seems to differ very little from *I. suaveolens (I. mellita)*. Living plants recently collected by T. Baytop have flowered in my garden and they show that the bracts are both keeled as in *I. suaveolens.*

I. griffithii Baker. This central asiatic species is very little known and the information is based only on rather poor herbarium material. It is a fairly short species, about 20–35cm in height, with a two-flowered unbranched stem. The leaves are about 1–2·5cm wide, the outer ones in the fan slightly curved, and the bracts look as if they might have been wholly green, not transparent. The 5·5–6cm diameter flowers are purple, probably with a dense white beard on the falls, and they have a perianth tube 4–5cm in length. The original collection was made in eastern Afghanistan and it seems odd that it has not been recollected in this relatively well-known part of the country. The long perianth tube and long rather narrow green bracts give this species a distinctive appearance.

I. guertleri (Prodan) Prodan. A hybrid between *I. pumila* and *I. aphylla.*

I. hungarica Waldst. & Kit. A synonym of *I. aphylla.*

I. illyrica Tommasini. This is mentioned under *I. pallida*. It is somewhat intermediate between subsp. *pallida* and subsp. *cengialtii.*

I. imbricata Lindley (Syn. *I. obtusifolia; I. sulphurea; I. talischii*). This is a stout but compact plant, normally 30–60cm in height when in flower with short, stiff, usually straight, grey-green leaves about 2–3cm wide. The main stem is branched, with the lateral flowers only shortly stemmed or sessile. A prominent feature is provided by the much-inflated bracts which are thinly membranous but pale green, not dry and transparent except at the extreme tip. In all the plants I have seen in the wild and cultivation the flowers have been a dull or pale yellow with a narrow, darker-yellow beard and usually some brownish veining on the haft of the falls. It is however, noted by Rodionenko that bluish-flowered forms can be found. The flower diameter is about 7–9cm, large for the size of the plant and since it is possible for there to be up to three flowers open at any one time on the stem, it can be a very handsome plant.

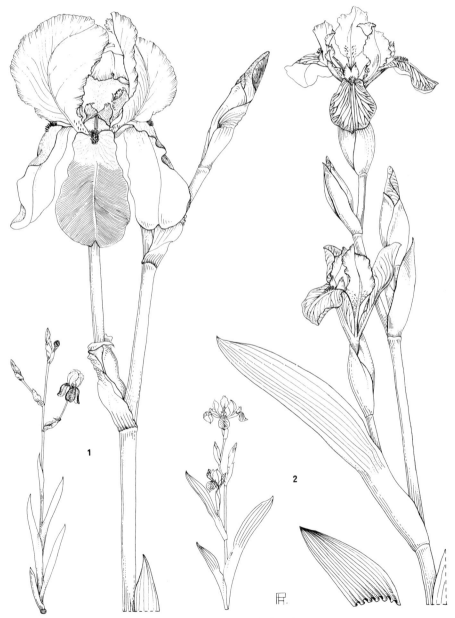

Subgenus Iris, Section Iris: 1 *I. kashmiriana*; 2 *I. variegata*

I. imbricata occurs in damp rocky places or mountain meadows at 1400–3000 metres in the Elburz mountains of Iran, from the central part of the range westwards and northwards around the Caspian through the Talysh mountains into eastern Transcaucasia. It is obviously related to *I. taochia* from north-eastern Turkey and I have made comments about the distinguishing features under the latter species.

An unforgettable sight for me when I visited Iran one year was a scattering of clumps of this iris amid red oriental poppies on the scree slopes of the high Elburz.

I. italica Parl. A synonym of *I. lutescens.*

I. junonia Schott & Kotschy. See under *I. germanica.*

I. kashmiriana Baker (Syn. *I. bartonii*) Although this tall bearded iris looks fairly similar to several others in general form, it may be recognised by the very long (up to 11cm) narrow green bracts. It has robust stems 75–125cm in height with one or two branches and the distinctly ribbed leaves are straight, pale grey-green, 2–3cm wide and up to 60cm long. The deliciously scented flowers are about 10–12cm in diameter and are either white or pale lilac-blue, with a dense beard of white, yellow-tipped hairs. There is a similar but smaller beard on the lower part of the standards. In its white form it can be separated by its stemmed lateral flowers from *I. albicans,* in which they are sessile, and from white forms of *I. germanica* (e.g. 'Florentina') by having long wholly green bracts. The *I. germanica* variants have shorter tubbier bracts which are transparent or brown in the upper half. *I. kashmiriana* is probably a wild species from Kashmir, also recorded in Afghanistan and Iran but only as a garden escape there. It is an attractive plant, apparently quite hardy in southern Britain and all but the coldest states of North America, rapidly forming large clumps.

I. kochii Kerner ex Stapf. This is regarded by Dykes as of probable hybrid origin, possibly between *I. germanica* and *I. pallida* subsp. *cengialtii.* Its origin (near Trieste) and characters certainly suggest that this might be so. I have grown the plant and my observations confirm those of Dykes that it is a shorter plant than *I. germanica,* about 45cm, with one or two branches on the stem. The flowers have the falls and standards similar in colour, a rather deep bluish-purple with insignificant brown-purple veining on the haft of the falls.

I. lepida Heuffel. A synonym of *I. variegata.*

I. leucographa Kerner. A synonym of *I. variegata.*

I. lisbonensis Dykes. A synonym of *I. subbiflora* var. *lisbonensis.*

I. × lurida Aiton. This is a name covering the hybrids between *I. pallida* and *I. variegata,* of which two well-known ones are *I. × squalens* and *I. × sambucina.* In north-west Yugoslavia and northern Italy the two parent species meet and produce a wide range of intermediates. Some of these are cultivated and some have become naturalized in other countries.

I. lutescens Lam. (Syn. *I. chamaeiris*). Unfortunately, because of the international rule of priority of names, this popular garden plant has to be called *I. lutescens* rather than by its much better-known synonym *I. chamaeiris*. It is a very variable plant with leaves from 0·5–2·5cm wide and up to 30cm long, not or scarcely curved. The height also varies considerably between 5cm and 30cm at flowering time. Unlike *I. pumila* the bracts are not tubular and closely sheathing the perianth tube but are shorter (3·5–5·5cm long) and more loosely arranged. The one or two 6–7cm diameter flowers have a perianth tube only 2–3cm long and may be yellow, violet or a mixture of the two colours, or occasionally white. The beard is yellow, unlike that of the similar *I. subbiflora* in which it is white or purplish, at least in the upper part on the blade of the falls. In the latter species the perianth tube is longer, up to 5cm.

   I. lutescens occurs in dryish, grassy or rocky places, sometimes beneath pine trees, in north-eastern Spain, southern France and Italy and it flowers in March or April. The great variation has inevitably led to a profusion of names so that apart from *I. chamaeiris* there are as synonyms *I. statellae*, *I. olbiensis*, *I. virescens*, *I. italica* and *I. erratica*.

I. majoricensis Barc. Possibly a synonym of *I. albicans*.

I. mangaliae Prodan. A synonym of *I. variegata*.

I. marsica I. Ricci & Colasante. This is a recently described species from the Abruzzi National Park in Italy and one I have not yet seen in the living or dried state. It is about 70–80cm in height and has broad arching leaves up to 5cm wide. The three or four flowers, on a branched stem, are produced from bracts 5·5–6·5cm long, exceeding the perianth tube which is about 3·5cm long. Each flower is about 8–9cm in diameter and resembles that of *I. pallida* subsp. *cengialtii*. It can have a yellow or white beard, the rest of the flower being a variable shade of violet with veining on the haft of the falls.

   I. marsica is apparently a wild species occurring in the Central Appenines on Mount Marsica where it grows on limestone hills at 1100–1700 metres altitude. It flowers in May or June.

I. mellita Janka. A synonym of *I. suaveolens*.

I. melzeri Prodan. A synonym of *I. aphylla*.

I. mesopotamica Dykes. See under *I. germanica*.

I. × neglecta Hornem. A hybrid between *I. variegata* and *I. pallida*.

I. nudicaulis Lam. A synonym of *I. aphylla*.

I. × nyaradyana Prodan. A hybrid between *I. variegata* and *I. pallida*, according to *Flora Europaea*, Volume 5.

I. obtusifolia Baker. A synonym of *I. imbricata*.

I. olbiensis Hénon. A synonym of *I. lutescens*.

I. pallida Lam. subsp. **pallida.** (Syn. *I. dalmatica* Hort.) The really distinctive feature of this lovely bearded iris is the wholly silvery papery appearance of the bracts, quite unlike those of any other species. It is a tall plant with branched stems up to 1·2 metres in height and grey-green leaves 1–4cm wide, usually about half as tall as the stems. The two to six large, deliciously scented flowers are about 9–11cm in diameter and are soft lilac-blue with a yellow beard.

I. pallida subsp. *pallida* grows naturally in western Yugoslavia, particularly in the Adriatic region, and is probably naturalized in other adjacent areas. I have seen it in great quantity as far south as Kotor where it flowers in May on limestone hillsides mixed with the rather rare yellow-flowered shrub of the Leguminosae, *Petteria ramentacea*. The plant grown in gardens as var. *dalmatica* does not seem to differ appreciably from typical wild plants.

I. pallida var. *illyrica* (Tommasini) Dykes (Syn. *I. illyrica* Tommasini) is a shorter plant than the typical *I. pallida* and resembles subsp. *cengialtii* in this respect, but its leaf and seed characters suggest that it is better placed with subsp. *pallida*.

I. pallida subsp. **cengialtii** (Ambr.) Foster (Syn. *I. cengialtii* Ambr.). This variant is shorter than subsp. *pallida*, usually about 30–45cm in height, and has greener leaves. The bracts are wholly papery and semi-transparent but are brownish rather than silver, and the flower is of a deep, slightly bluish-purple. The beard usually consists of white hairs tipped with yellow or orange. I have a plant which was collected by myself and Chris Grey-Wilson in which the beard hairs are wholly white. This may in fact be better referred to as subsp. *pallida* var. *illyrica* since it has glaucous, not green, leaves.

All the variants of *I. pallida* are easily grown garden plants for the open border.

I. panormitana Todaro. A synonym of *I. pseudopumila*.

I. perrieri Simonet ex P. Fournier. This is given as a synonym of *I. aphylla* by *Flora Europaea*, Volume 5. It grows in the French Alps and is said to be about 15–30cm in height with one to three bluish-violet flowers.

I. polonica Blocki ex Asch. & Graebn. A synonym of *I. aphylla*.

I. pseudopumila Tineo. (Syn. *I. panormitana*) As its name suggests, this species might be mistaken for a form of *I. pumila*. In general it has longer stems, at least 3cm, so that including the 5–7·5cm long perianth tube, the total height at flowering time to the top of the flowers is 15–25cm. The grey-green leaves which are present during the winter, unlike those of *I. pumila*, are about 1–1·5cm wide and only slightly curved. Like *I. pumila* the bracts, which are however somewhat more swollen, close in at the apex to sheath the tube closely and are very long, sometimes as much as 10–12cm. The flowers are solitary and about 6–8cm in diameter, varying in colour widely from purple to violet and white or yellow, sometimes bicoloured. It is a native of Sicily, Malta, Gozo and south-east Italy and is also recorded on the Adriatic coast of Yugoslavia. The habitat is in rocky or dry grassy places below 350 metres and it flowers in March or April.

Although this is an easily grown and useful plant for the rock garden, I find the leaves, and general appearance, rather coarse for the size of the plant.

I. × pseudopumilioides (Prodan) Prodan. A hybrid between *I. pumila* and *I. aphylla*.

I. pumila Linn. (Syn. *I. aequiloba; I. taurica*) Undoubtedly this is the most well known of the smaller pogon irises and is much-used in hybridizing as one of the parents of many of the popular dwarf bearded cultivars. The wild plant has intrigued botanists and cytologists for decades and some excellent work by Mitra and Randolph has led to some quite convincing ideas about its origin as a species. On the basis of chromosome structure and geography they showed that *I. pumila* most probably arose as a hybrid between *I. attica* and *I. pseudopumila*, the overlap in their distributions being in the Adriatic region of Yugoslavia. Having established itself, the species then spread north eastwards into eastern Europe and into Russia as far as the Ural mountains, the limit of its present distribution. The work is published in considerable detail in the *American Journal of Botany* **46**, pages 93–102 (1959).

I. pumila is now firmly established as a species in its own right and like most other widespread bearded irises shows considerable variation. I have taken it in a broad sense here, but some botanists, notably the Roumanian J. Prodan, have given many of the variants names, at various taxonomic levels. For those who wish to follow up with some extra studies of *I. pumila*, Prodan's work can be found in *Buletinul Gradinii Botanice si al Muzeului Botanic dela Universitatea din Cluj,* Volume **14** (1934); **15** (1935) and **25** (1946), and in the *Flora Republicii Socialiste România* **11** (1966).

I. pumila is a dwarf plant with stems usually not more than 1cm long, up to the base of the bracts. Since the perianth tube may be 5–10cm long, the total height to the top of the flower can be 10–15cm. Sometimes taller-stemmed variants occur, especially in the eastern end of its range—for example var. *elongata* of Lipsky—which can apparently have stems up to 12cm long. The nearly straight greyish-green leaves vary from about 10–15cm long and 7–15mm wide. There is normally one flower, the tube of which is closely sheathed by two long bracts, 5–10cm long, which are rounded and not keeled at all, or the outer one only slightly keeled; the outer one is usually green and the inner, more membranous one is much paler. The flowers are scented, about 5–6cm in diameter and may be yellow, purple or blue with a yellow or bluish beard on the falls.

I. pumila occurs wild in central and northern Yugoslavia and Austria eastwards through Bulgaria, Czechoslovakia, Hungary and Roumania to Russia where it reaches to the Urals. It is a plant of dryish grassy places. In winter the leaves of *I. pumila* die away whereas those of the similar *I. pseudopumila* do not.

As with other small pogon irises I do not find this species an easy plant to keep for long. It does well for a year or two and then dwindles away. The best chance of success seems to be in a raised bed which is lime-rich and well supplied with nutrients which are not too high in nitrogen content.

Soft growth caused by too much nitrogen, especially in the late summer, usually leads to heavy losses in the winter.

I. reginae Horvat. This is a variant of *I. variegata*.

I. reichenbachii Heuffel (Syn. *I. balkana; I. bosniaca; I. skorpilii*). This species closely resembles *I. suaveolens* but is usually a little taller and has broader leaves and a shorter perianth tube. It grows 10–30cm in height with leaves about 1–1·5cm wide. The bracts are both keeled, not rounded as in *I. pumila*. There are one or two terminal flowers and sometimes a lateral branch also, with a stemless flower. These are about 5–6·5cm in diameter and may be dull yellow, dirty purple or violet, often veined darker and with a beard of yellow or purple-tipped white hairs. The perianth tube is usually about 1·5–2·5cm long and the standards are rather large, often with a wider blade than the falls.

   I. reichenbachii grows in northern Greece, certainly as far south as Mount Olympus where I have seen it in the yellow form, in Bulgaria, central and southern Yugoslavia and Roumania. It grows in the mountains, usually in grassy places or in scrub. Although an easy plant to grow in open sunny situations it is not one of the most attractive species, the flowers often having a rather dowdy appearance.

   I. balkana is probably only a purple-flowered form of *I. reichenbachii*.

I. ricardii Hort. A synonym of *I. mesopotamica*.

I. rosaliae Prodan. A hybrid between *I. variegata* and *I. pallida*.

I. × rothschildii Degen. A hybrid between *I. variegata* and *I. pallida*.

I. rubromarginata Baker. A synonym of *I. suaveolens*.

I. rudskyi Horvat. A synonym of *I. variegata*.

I. × sambucina Linn. A hybrid between *I. variegata* and *I. pallida*.

I. scariosa Willd. ex Link. (Syn. *I. glaucescens; I. eulefeldii*). This is a dwarf clump-forming iris which grows about 10–20cm in height and has the rhizome furnished with short brown fibres. The leaves are slightly curved and are pale greyish-green. Each stem produces two flowers from the rather narrow thin-textured bracts which do not closely sheath the perianth tube. Compared with that of *I. pumila*, the tube is rather short, about 3cm in length. The flowers are about 4–5cm in diameter and are reported as being reddish-violet, light to dark violet, blue to nearly white or yellow, although the yellow forms appear to be more uncommon.

   I. scariosa is an inhabitant of dryish steppe country at altitudes of up to 3000 metres, often in saline soils, where it flowers in May. It is a Russian endemic, but even so has a very wide distribution from the Ural region east to the central asiatic Tien Shan mountains where it is a somewhat larger plant.

I. schachtii Markgraf. This comparative newcomer to cultivation was named after Wilhelm Schacht the great German gardener, until recently Curator of the Munich Botanic Gardens. It has now been collected many times in Turkey and it is clear that it is a quite widespread plant in central Anatolia. Needless to say, coming from that region, it is extremely hardy but does require plenty of sun in summer, a problem when brought into more northerly gardens. It is a short plant, about 10–30cm in height at flowering time with grey-green leaves up to 1·5cm in width. The stems often have one to three lateral branches, the lowest of which may be up to 5cm long. The thin-textured bracts are broad and slightly inflated, green or tinted with purple, and transparent towards the margins and apex. Various colour forms exist, but the species is commonly yellow or purple-flowered. In the yellow forms there is usually some greenish or brownish veining on the haft of the falls and occasionally the whole blade is a brownish or purplish colour. The beard is yellow in the yellow-flowered forms and pale blue or yellow in the purple forms. The flowers are of good size, usually about 5–6cm in diameter, and often there is more than one out at a time on each inflorescence.

I. schachtii was first gathered near Ankara but now its range is known to stretch from Eskişehir province eastwards to Kayseri and Sivas provinces. It grows in dry open rocky situations in scrub or in open woods at altitudes of 400–1700 metres and it flowers in May.

Occasionally the lowest branch of the inflorescence starts at a point below the middle of the main axis and this has given rise to the belief that *I. aphylla* occurs in Turkey. The names *I. flavescens* and *I. lutescens* have also been mentioned in connection with the Turkish flora and I believe that these records too can be assigned to *I. schachtii*.

The narrow leaves, 1·5cm wide at most, separate *I. schachtii* from the more gross *I. taochia* which occurs farther to the east in Turkey.

I. sicula Todaro. Either a synonym of *I. pallida*, or possibly a hybrid between *I. pallida* and *I. germanica*.

I. skorpilii Velen. A synonym of *I. reichenbachii*.

I. × squalens Linn. A hybrid between *I. variegata* and *I. germanica*.

I. statellae Todaro. A synonym of *I. lutescens*.

I. suaveolens Boissier & Reuter. (Syn. *I. mellita; I. rubromarginata; I. glockiana*). This delightful dwarf species has long been familiar in gardens as *I. mellita*, although it is in fact the same as *I. suaveolens*, which is the older and therefore correct name. It is about 8–15cm in height when in flower and has small fans of curved leaves, each 4–10cm wide. The one or two flowers are produced from sharply keeled bracts and they are usually about 4·5–5·5cm in diameter. As with many of the dwarf bearded iris species there are several colour forms in yellow, purple, yellow with a brownish or purplish blade to the falls, or dull brownish-purple throughout. The beard is yellow, or sometimes bluish in the purple-flowered forms. *I. suaveolens* grows wild in eastern Bulgaria and north-western Turkey, both in the European and Asiatic parts. It is also in the Black Sea region of south-east Roumania and possibly in southern Yugoslavia although I am not sure if the plant I saw once near Skopje was this or a short-stemmed *I. reichenbachii*. It is especially common on the hills around Üsküdar on the Bosphorus opposite Istanbul where I have seen it in great quantity in several colour forms. Occasionally one can find forms in which the leaves are edged with purple, these variants being responsible for the name *I. rubromarginata*.

The keeled bracts separate this from *I. attica* and *I. pumila*, in which both bracts are smooth or sometimes the outer one is slightly keeled. Moreover the perianth tube of *I. suaveolens* is usually 4–4·5cm long, while in the other two species it can reach 10cm, and is not less than 5cm long. *I. reichenbachii* also possesses keeled bracts but this is usually a taller plant about 10–30cm in height with leaves up to 1·5cm wide and with a perianth tube usually only 1·5–2·5cm long.

I. subbiflora Brot. This grows to 25–40cm in height and has nearly straight leaves about 0·5–2·5cm wide, with some reduced leaves borne on the stem as well. The one or two flowers are deep violet, about 7–8cm in diameter, and have a tube 3·5–5cm long. The beard hairs may be violet or white, but on the lower part of the haft of the falls are usually yellowish. Although it is rather similar to *I. lutescens (I. chamaeiris)*, which may have violet forms, the longer perianth tube and the colour of the beard distinguish it from its more easterly occurring relative.

I. subbiflora is a native of Portugal and adjacent south-western Spain where it grows in open rocky places or in sparse grass or light scrub. It is a good garden plant and seems to be as hardy as any of the tall bearded iris cultivars.

I. subbiflora var. *lisbonensis* (Dykes) Dykes. This is a variant in which the stems are more or less leafless and the spathes are green, not purple-tinged as in var. *subbiflora*.

I. sulphurea Koch. A synonym of *I. imbricata*.

I. talischii M. Foster ex Spreng. A synonym of *I. imbricata*.

I. taochia Woronow ex Grossheim. This interesting plant occurs locally in the north-eastern part of Turkey. It has only in the last decade been introduced into cultivation and I am indebted to T. Baytop for giving me rhizomes, and for taking me to see wild populations of it in Turkey. It is proving to be a good garden plant in Surrey, although the flower colour is perhaps not very exciting.

I. taochia is a rather compact species with broad leaves, the stems being only 15–30cm tall with one to three short branches, and the leaves 1·5–2·5cm wide. The leaves are often as tall as the stems so that the flowers are carried amid them, giving the appearance of a very leafy plant. There is considerable variation in flower colour from pale yellow to almost gold and from dirty purple to deep violet. The beard consists of yellow-tipped white hairs, or wholly deep yellow hairs, and on the haft of the falls there is brownish, brownish-purple or violet veining on a white or yellow ground. The bracts are rather inflated, membranous and green, similar to those of *I. imbricata*.

I. taochia grows in rocky places at 1500–1700 metres to the north-east of Erzurum in Turkey where it flowers in May or June. It apparently occurs only on basalt rocks and, in the locality where I have seen it, grows in considerable quantity in large circular clumps up to 1 metre across.

Although *I. imbricata* and *I. taochia* are similar there do seem to be some good distinguishing features. In the former the flowers are about 7–9cm in diameter and always yellow (bluish forms are reported but I have not seen any convincing material; yellow-flowered plants often dry bluish in herbarium specimens). The flower stems of *I. imbricata* are about 30–60cm in height. *I. taochia* is a shorter plant, about 30cm in the tallest examples I have measured, the flowers are yellow or purple (white is also recorded) and only about 5–6cm in diameter. They are, I think, also distinguishable on leaf characters, those of *I. imbricata* being abruptly narrowed to a short pointed tip and those of *I. taochia* tapering more gradually to a long point. Moreover I have observed that in *I. taochia* they often have a longitudinal twist while in *I. imbricata* they are not twisted but are held in a strictly flat plane.

I. taurica Lodd. A synonym of *I. pumila*.

I. timofejewii Woronow. This is a rather dwarf species, usually about 10–25cm in height, with narrow (about 5mm wide) sickle-shaped greyish leaves. The bracts are sharply keeled and subtend two flowers, each about 4·5–6cm in diameter, of a deep reddish-violet. The falls have a white beard, usually with purple tips to the hairs. The perianth tube is about 4–5cm long. It is a native of Dagestan in the eastern Caucasian region of the USSR and is described as being a plant of sunny rocky slopes in lime-rich soil. I have grown and flowered it in Surrey but it does not seem to persist for long. The answer may be to include it in a bulb frame with Oncocyclus and juno irises, for Rodionenko observes that it requires plenty of sunshine and good drainage.

It is very similar to some forms of *I. reichenbachii* but has a longer perianth tube and narrower leaves.

I. trojana Kerner ex Stapf. See under *I. germanica*.

I. varbossiana Maly. This is a little-known plant, originating from Bosnia and said by Hayek to be a variety of *I. germanica*.

I. variegata Linn. (Syn. *I. rudskyi; I. reginae; I. lepida; I. leucographa; I. mangaliae; I. dragalz; I. virescens*). This rather dramatically coloured bearded iris is one of the most distinct. It is of medium height, usually about 20–45cm when in flower, and has strong, deep green, slightly curved, leaves which are very noticeably ribbed and about 1–3cm wide. The stems are branched and carry three to six unscented flowers which are subtended by inflated green or purple-tinged bracts. Each flower is 5–7cm in diameter and is conspicuously bicoloured, the standards and style branches pale or dull yellow and the falls heavily veined reddish-brown or violet on a whitish or pale yellow ground. The beard is yellow. Sometimes the veining on the falls runs together into a complete stain of brown-purple on the blade producing a very striking contrast with the yellow standards. There are also on record forms with a whitish ground colour to the falls and standards which are both violet-veined.

One such form, which also differs from 'typical' *I. variegata* in having a slightly longer perianth tube, was described as *I. reginae* by I. & M. Horvat in 1947. Original material of this is still in cultivation in Britain, said to have been collected in Yugoslavia. Another pleasant form in yellow with a brownish-red blade to the falls was described by the same botanists as *I. rudskyi* and this too can still be seen in some specialist iris collections in England.

I. variegata occurs wild in central and eastern Europe, from southern Germany through Austria, north-eastern Italy, Yugoslavia, Hungary, Czechoslovakia, Roumania and Bulgaria eastwards to the western Ukraine. It is a plant of light woodland, scrub and open stony places, flowering in May or June.

I. virescens Delarb. A synonym of *I. variegata*.

Iris sp. Turkey. This is possibly a distinct species, as yet un-named. It is one which I have grown for about 14 years, and have collected in the wild, but in this section it seems to me to be unwise to describe yet another species before a thorough study of the group has been completed.

It is about 30–50cm in height with one to three branches to the stem, the lowest usually only a few centimetres in length. The leaves are about 1–2cm wide and rather short, usually 10–20cm. A characteristic feature seems to be the wholly purple-tinged thin papery bracts which are rather short and somewhat inflated. The scented flowers are about 6–7·5cm in diameter and may be either a pale pearly blue suffused with darker blue in the centre of the falls, or pale yellow suffused with pale purple on the lamina. The haft of the falls is veined with dull violet or brown and the beard hairs are white tipped with yellow.

This variable plant has been recorded by several people in western and south-western Turkey and one particularly attractive form is that collected by Asuman Baytop which has pale icy-blue flowers and purple bracts.

In addition to the species described in this section there are other populations of very variable irises, especially in Turkey, which are in need of careful investigations in the field. In some instances they cover within one area the whole range of colours through white, cream, yellow, blue, lilac and purple and vary also in stature and bract characters.

1B   Subgenus Iris, Section Psammiris

This small group, *I. humilis (I. arenaria)* and its allies, is closely related to the Regelia group but has been included with the bearded (pogon) irises by some authorities. They are separated from the pogons (Section Iris) by having seeds with a fleshy aril and thus must be more associated with the arillate bearded species of Sections Regelia, Hexapogon, Oncocyclus and Pseudoregelia. They are distinguished from the first two groups by having a beard only on the outer segments (falls) and from Pseudoregelia in the rhizome characters. In Psammiris the rhizomes are usually stoloniferous or at least shortly-creeping, whereas in Pseudoregelia the plants are strictly clump forming with no sideways movement of the growing point over several seasons. *I. potaninii* however is a debatable case for it is more clump-forming. John T. Taylor in his classification of these arillate-seeded irises also points out that the species of Pseudoregelia Section have flowers which are characteristically spotted and blotched (as in *I. tigridia* and *I. kamaonensis*). From Section Oncocyclus, the Psammiris species can be distinguished by having generally smaller flowers, usually more than one in each set of spathes, and coloured yellow in all the species (sometimes lavender in *I. potaninii*). The Oncocyclus all have solitary large flowers in which yellow is an unusual colour, tending to be an occasional mutation. Furthermore, the aril on the seeds of Oncocyclus is very large whereas in Psammiris the aril is much smaller than the body of the seed.

The Psammiris iris species

I. arenaria Waldst. & Kit. A synonym of *I. humilis*.

I. bloudowii Bunge. This is allied to *I. humilis* and the two might be variants of one species, although there are certain characteristic features. It grows 15–35cm in height and has strong leaves up to 30cm long and 7–13mm wide. These are erect or slightly curved and because of their larger size make the whole plant look much leafier than *I. humilis*. The rhizome is non-stoloniferous and is rather stouter than that of *I. humilis*, and it has many brown fibres round the growing point. The spathe valves or bracts are very inflated and contain two or three yellow flowers which have a purple or brownish stain on the lower part of the segments. The falls have a yellow beard like *I. humilis* but the whole flower is rather larger, usually about 5cm in diameter. It is a native of Siberia and Russian Central Asia in the Tien Shan mountains, east to Mongolia and the far eastern provinces of the USSR. It is possibly also in China. The habitat is mountain wood margins, or scrub or alpine meadows at about 1500–2000m, flowering in May or June in the wild.

Although at present I do not grow this plant I have done so in the past and it appears to be somewhat less tricky to keep than *I. humilis*.

I. flavissima Pallas. A synonym of *I. humilis*.

I. humilis Georgi (Syn. *I. flavissima*, *I. arenaria*, *I. flavissima* subsp. *stolonifera*, *I. pineticola*). There has been a great deal of ink and paper used in the debate about the name of this little iris and no doubt there will be more. Part of the problem lies in deciding whether there is one or more species involved; if more than one, then the question arises as to how to apply the many names available. I can see no reason for 'splitting' this and prefer to regard it as a single, rather widespread, and not particularly variable, species. It is a dwarf plant, about 5–25cm in height with erect leaves about 2–7mm wide, which tend to curve inwards at the apex. The flower stems carry one or two (rarely three) flowers which are about 3–4cm in diameter and are yellow with an orange beard. The blade of the falls spreads out horizontally rather than deflexing so that the flower has a slightly flattish and solid appearance, because of the broad rounded apex of the falls. The standards are shorter than the falls and the style crests are also short so that the whole flower has a squat appearance. *I. humilis* occurs in eastern Europe, in central Czechoslovakia, Austria, Hungary, Roumania and south-east Russia, and in Asiatic Russia in Siberia right across to the far eastern provinces and Mongolia. It is an inhabitant of sandy steppes and stony slopes at about 200–1500 metres and flowers in April to June. Attempts have been made to name the variants of this species and for anyone requiring more information there is an account by K. H. Ugrinsky in Fedde's *Repert. Spec. Nov.*, Beihefte **14** (1922). It does seem that there is a tendency for the European material to be more creeping in its habit, with slender stolons and with narrower leaves than the Asiatic plants, which are inclined to be more tufted with broader leaves. However, looking at a wide range of herbarium specimens there seems to be no obvious distinction between these characters and I find the division scarcely convincing.

I do not find it an easy plant to grow.

I. mandschurica Maxim. A little-known species, obviously related to *I. bloudowii*. I have seen only dried material and this shows a plant about 10–15cm in height with a stout, non-stoloniferous rhizome, thicker than that of *I. bloudowii*. The leaves are more or less straight and about 8–10mm wide. The flowers are yellow and also look very similar to those of *I. bloudowii* as far as I can tell. It grows in the far east of Russia in the Ussuri region and in Manchuria and possibly Korea. Flowering in the wild is in May but I have no data about its wild habitat.

I. potaninii Maxim. A beautiful little plant, unfortunately little-known in cultivation if indeed it is represented at all. The rhizome has strong fleshy roots and is compact, spreading sideways only slightly so that the plant forms dense clumps. A very curious and characteristic feature is the mass of short curled fibres which remain attached to the rhizome and stick up as a fibrous clump above the soil surface. The leaves are straight and erect, only 5–10cm long and 2–3mm wide. The flowers are held at the same level or just above the leaves and are solitary and about 3–4cm in diameter. There are two colour forms, yellow or lavender-purple, both with a narrow yellowish or whitish beard. There is some resemblance between the purple-flowered form and *Iris tigridia*, a member of the Section Pseudoregelia, but they can be distinguished by the flowers of the latter being conspicuously blotched darker, the lack of curly fibres on the rhizome and by the leaf apex which in *I. potaninii* is fairly obtuse and in *I. tigridia* tapered to a fine point.

I. potaninii occurs wild in mountainous western China and in central and eastern Tibet. It flowers in May and June at altitudes of 3500 to nearly 6000 metres, and inhabits open rocky slopes, sometimes in spruce forests.

1C   Subgenus Iris, Section Oncocyclus

This group of rhizomatous species contains some of the most bizarre irises of all, not to mention some of the most infuriatingly difficult, as far as cultivation in

Britain is concerned. The 'Oncos' are on the whole xerophytic plants occurring from central Turkey to the Caucasus, south to the Sinai peninsular and east to the Kopet Dag mountains in north-eastern Iran. In these areas they are subjected to a dry, often hot, summer climate, usually coupled with a relatively dry cool winter. The winter months may be spent beneath a thick layer of snow—this applies to most of the species from Caucasia, Turkey and Iran—and the plants are more or less in a state of dormancy. The more southerly occurring species are not snow-covered in winter but are cool and dryish and normally do not make much growth until the early spring, so that even those inhabiting southern Israel do not flower before March or April. In the more northerly mountains flowering may be as late as May or June after the snow has melted. It can be fairly safely said that all Oncocyclus species have a long dryish dormant period in summer, usually warm or hot, depending upon the altitude. In the autumn, with the soil temperature still reasonably high, there will normally be some rain, although in northern mountain districts this rapidly changes to snow. During this humid period, root growth at least will begin, if not some leaf growth also. Eventually all growth will be slowed right down by the falling temperatures and there will be little activity until springtime. A point to bear in mind is that although the rhizomes of the irises are near or at the surface, their roots are mostly long and reach well down into the soil where there is a certain amount of moisture and, in winter, relative warmth compared to the temperature of the surface. Obviously these comments are generalizations, for each species has an ecological niche of its own, in most cases not studied in any detail. One would like to see, for example, soil and air temperature figures throughout the year, precipitation measurements, snow cover, soil types, precise habitats etc., recorded for even just a few of the species to give some idea of the conditions they encounter during one complete year.

Obviously those living in Britain and the damper northern parts of the United States are off to a bad start with our erratic climate. Our summers might be sunless and wet and the winters warm and muggy! We can provide some form of protection and control of the environment, and some of the ways in which we can increase the chances of success are outlined below. However those regions such as the central and south-eastern United States—which have a more continental type of climate with drier, more well-defined summers—are well suited to the needs of this group of irises.

The Oncocyclus iris group is fairly readily defined botanically in being rhizomatous, with the stems having only one flower which has bearded falls and by having the seeds furnished with a conspicuous, often whitish, aril which is usually nearly as large as the body of the seed itself. The pogon or bearded irises have no aril on the seeds, and the related Regelia species, which do have arillate seeds, can generally be distinguished by having two flowers per stem and a beard on the standards as well as the falls. Additionally, there are cytological reasons for separating these more easterly-occurring Regelias from the Oncocyclus species.

Cultivation

From the notes given above it is clear that one essential ingredient for the successful culture of Oncocyclus irises is the provision of some form of protection from excess summer moisture. With such protection the rhizomes can go into a state of proper dormancy for a lengthy period. Good drainage must be given, for many of the species grow naturally on hills where there is a rapid 'run-off' of excess rain. Plenty of light and air are essential since all the species grow in areas with little competition or shade from other plants—they are normally plants of open airy sites in regions of high light intensity. The natural soils, although perhaps very dry at times, are not necessarily poor and there is usually plenty of nourishment and of course unlimited root-run, even in rocky places.

So, what does all this mean? In Britain and those parts of the United States with a similar climate where Oncocyclus irises will not flourish in the open garden, we can build raised beds for drainage purposes, they can be filled with a gritty soil which is rich in certain fertilizers and these beds can be provided with glass or plastic covers for protection from excess wet in winter and summer. A bulb frame into which the irises can be planted is probably the ideal method of cultivation, for it provides an unrestricted area for growth. In pots the precise conditions may be under even more direct control, but the roots are very restricted and watering, feeding and frequent repotting becomes a much more laborious process.

Assuming a bulb frame has been built along the lines suggested in the general chapter on cultivation (page 7) the annual treatment of Oncocyclus (and Regelia) irises is as follows. Firstly, planting the rhizomes must take place not later than October if possible; they should be covered with a 3–4cm layer of coarse chippings. Water is given during October and then withheld until the following February or March depending upon the mildness of the season. Through the month following planting the aim is to encourage root growth with a warm moist soil but keeping top growth to a minimum so that as much air as possible is allowed through the sides of the frame. This applies subsequently, right through the winter unless a really low-temperature period is encountered in which case the sides of the frame are closed to prevent the soil freezing to too great a depth. In early spring, watering can be started again, with maximum ventilation and feeding using a 'low nitrogen' mixture. Any proprietary brand will do, either as a liquid feed or in granular form, scattered on the surface so that the nutrients are washed down slowly with each watering. Flowering in Britain normally takes place in April or May and usually watering can be ceased from about June onwards. Obviously if seeds are produced, water should not be withheld until the capsules are ready to burst open. The foliage should then wither and die away for the rest of the summer, at the end of which division and replanting can take place. There is no need to remove the lights at all through the year if they are of the type with sliding sides. In fact this type is preferable since there is then always a cover over the plants to prevent heavy rain damaging the flowers. There is no reason why the lights should not be removed if desired during the early spring and left off until the drying-out period in summer. Oncocyclus irises are very susceptible to virus diseases, and aphids should be kept under control at all times as a precautionary measure since they are the carriers of these devastating virus diseases.

Cultivation in pots is essentially the same as in bulb frames as far as the year-round treatment is concerned. The plants are best housed in an alpine house, or at least somewhere where there is plenty of light and air, and the pots should be plunged in sand to keep the roots at as constant a temperature and humidity as possible. Fortnightly feeds are advisable since Oncocyclus irises, when growing well, are strong growers and have vigorous roots which will rapidly exhaust the nutrients in a pot of soil.

The Oncocyclus iris species

I. acutiloba C. A. Meyer. (Syn. *I. fominii, I. szovitsii*). A small species about 8–25cm in height with narrow sickle-shaped leaves only 2–6mm wide. The flowers are about 5–7cm in diameter, whitish, strongly veined and streaked brownish or greyish with a dark reddish-brown or blackish signal patch in the centre of the falls. The beard consists of sparse, long, purplish or brownish hairs. Both falls and standards are rather pointed, hence the name 'acutiloba'. According to Rodionenko, the great authority on Russian iris species, typical *I. acutiloba* has two dark spots on each fall, one in the centre (the signal patch) and one at the apex. In the subspecies *lineolata*, an Iranian plant, the flowers are furnished with only one spot on each fall (see below). *I. acutiloba* occurs in the Transcaucasus, to the north of the river Kura. Colour forms of *I. acutiloba* occur and at least five of these have been described by Russian botanists.

I. acutiloba subsp. **lineolata** (Trautv.) Mathew & Wendelbo (Syn. *I. helena, I. ewbankiana*). This subspecies, as mentioned above, has only one dark blotch on each fall and is distributed to the south of the river Kura in Transcaucasia, in north-western and north-eastern Iran and adjacent Turkmenistan. The plants growing in the last two areas were formerly called *I. ewbankiana* but I cannot find any consistent differences between these and plants from other areas of Iran.

This is a delightful plant, flowering in April and May on open stony steppe country and rocky mountainsides at 1500–3000 metres. It is clump-forming and very floriferous and in my view is one of the most graceful of all the Oncocyclus irises. In Transcaucasia there is a certain amount of variation and there are reported to be yellowish and creamy forms without signal spots or veining. Four variants have been described from this region. In cultivation it appears to be one of the easier species to grow, especially if planted into a bulb frame. Collections which we (D. Barter, S. Baker, D. Pyrcraft and B. Mathew) made on the Bowles Scholarship Expedition in 1963 grew exceedingly well in the frames at Wisley and soon made clumps up to 30cm in diameter. More recent collections by John Ingham are also proving satisfactory as alpine house pot plants, being of dwarf stature.

There are further comments about this species under *I. sprengeri* and *I. elizabethae*.

I. annae Grossheim. This is considered by Dr Rodionenko to be a hybrid which occurs spontaneously in Transcaucasia. He mentions that it has been found only three times and regards it only as an occasional non-permanent hybrid of no great consequence.

I. antilibanotica Dinsmore. This grows to about 25–40cm in height and is a large-flowered plant but with small, sickle-shaped leaves less than 10cm long. The flower is bicoloured with the falls purple-veined on a rich purple ground colour, without any prominent speckling. The standards are a much paler violet. The falls have a small blackish blotch in the centre and a yellow beard, the hairs of which are sometimes purple-tipped. It grows only in Syria on the mountains above Bludan where it flowers in May or June at about 2300 metres.

I. assadiana Chaudhary, Kirkwood & Weymouth. This smallish species, about 15cm tall, produces stolons up to 12cm long from the small rhizomes. The leaves are strongly curved and less than 1cm wide, and usually only about 5–10cm long. The flowers are 5·6–7·5cm in diameter and are of a uniform deep sombre maroon, or dark purple to blackish, with darker veins and a velvety signal patch on the falls. The beard consists of a mixture of short purple hairs and long yellow or dirty-whitish ones, these sometimes tipped with purple. The style branches are pale orange or reddish, streaked with purple. This plant is confined to the Syrian desert and is recorded in several places, such as Palmyra, Qaryatein and Hafar where it seems to grow in chalky hills or gravelly plains about 800–1000 metres, flowering in April. Chaudhary notes that W. R. Highwood saw white, yellow and pale forms around Qaryatein.

I. atrofusca Baker. This is one of the robust Israeli species with a stout rhizome and erect pale green leaves about 1cm wide. The large flowers are about 8–9cm in diameter and have dark purple-brown falls with a short dense beard of yellow hairs and a blackish signal patch in the centre. The standards are usually paler, more wine-coloured. Kenneth Bastow, who has grown and flowered many Oncocyclus irises, likens the overall colour to that of a freshly peeled horse chestnut, with a shiny, deep, rich appearance. It is a southerly species occurring in the deserts of Judea and Negev.

I. atropurpurea Baker. This rather showy 'blackish'-flowered species grows to about 15–25cm in height. The rhizomes often produce stolons so that it can make fine clumps with rather short sickle-like leaves. The flowers are about 8cm in diameter and are very dark or blackish-purple, usually with the standards paler than the falls. The beard hairs are short and yellow and the signal patch blackish. It occurs wild in the Gaza and Negev regions of Israel, and is a plant of low semi-desert sands where it flowers in March. There have been several varieties described, none of which I have seen in the living state.

I. atropurpurea var. *eggeri* Dinsmore. This was described as having a non-stoloniferous habit and brownish flowers and the leaves are said to be less curved.

I. atropurpurea var. *gileadensis* Dinsmore. A colour variation in which the flowers are said to be brown with red veins and dots. It is recognized as a separate species by Michael Awishai and comes from the Gilead region of the north-west Jordan, far removed from Gaza. In the one specimen I have seen the beard is deep violet.

I. atropurpurea var. *purpurea* Dinsmore. This was given specific status by Mouterde in *Nouvelle Flore du Liban et de la Syrie* as *I. bostrensis* and a description can be found under that name.

I. auranitica Dinsmore. A beautiful species up to 50cm in height with large yellow flowers 12–15cm in diameter which are spotted with brown, giving a rather bronze appearance overall. The signal patch can be maroon or reddish and the broad dense beard has yellow hairs, tipped with purple. It is restricted to the mountain block known as Djebel Druze in Syria where it grows in volcanic lava and flowers in May. Occasionally the signal patch and spotting are lacking altogether, and this form has been described as variety *unicolor* by Mouterde, and as forma *wilkiana* by Chaudhary.

In cultivation *I. auranitica* produces smaller flowers, as little as 8cm in diameter, and they are produced in April in Britain.

I. barnumae Baker and Foster. This delightful plant is one of the easier Oncocyclus irises to grow in Britain. It forms large clumps of curved, very grey-green leaves and in flower varies from 10–30cm in height. The flowers are about 7–8cm in diameter and are wholly deep purplish-violet with little veining on the segments. The beard is a narrow strip of yellow hairs, sometimes tipped with purple and sometimes whitish or creamy. The signal patch is not very obvious, being small and only slightly darker than the rest of the flower. *I. barnumae* occurs in eastern Turkey, north-eastern Iraq and adjacent areas of Iran and grows on stony or sandy hillsides at 1300–2300 metres. I have seen it in great quantity on the hills to the east of Lake Van where it grows near *I. paradoxa*. A plant collected by Hans Leep and Erich Pasche seems to be a hybrid between the two with rather intermediate characters.

I. barnumae subsp. *barnumae* forma *urmiensis* (Hoog) Mathew & Wendelbo. This is the yellow form of the species, slightly variable in its shade of yellow but always a beautiful clear colour, unspotted and unveined. The plant also varies somewhat in its stature and the dwarf, more compact forms are preferable to the rather lanky ones which seem to be the more common. The beard consists of fairly dense orange-yellow hairs and there is a slightly darker yellow signal patch. f. *urmiensis* takes its name from Lake Urmia (Rezaiyeh) in north-west Iran and it is a common plant on the mountains to the west of the lake. It also occurs, apparently more rarely, in south-eastern Turkey. The habitat is similar to that of the purple form, but I know of no instance where the two grow together.

I. barnumae subsp. **barnumae** forma **protonyma** (Stapf) Mathew & Wendelbo. This variant usually has brownish-purple falls and purple-violet standards and, in contrast to forma *barnumae*, the purple-black beard hairs are densely arranged and form a broad strip with a squared-off apex. It occurs in western Iran, especially around Lake Rezaiyeh. To the south of this, in the Hamadan and Sanandaj areas, it possibly hybridizes with *I. meda*, for there is a whole range of extraordinary plants with colours from the brown-gold of *I. meda* to dirty brown-purple or even lilac, and with beards varying from thin strips of yellow hairs to broad, dense, blackish bands. Paul Furse and Jim Archibald collected a range of these, some of which are very beautiful. Per Wendelbo also investigated them in the wild and it is still really not very clear as to what is the true status of the extraordinary populations from this region.

I. barnumae subsp. **demavendica** (Bornm.) Mathew & Wendelbo. This is usually a rather more robust plant than *I. barnumae* with broader more erect leaves, and it often has larger flowers of a more blue or bluish-violet shade. The beard too is a little different, being of white or creamy hairs confined to an even narrower strip. Geographically it is separated from the rest of the *I. barnumae* variants, for it occurs in the Elburz mountains of Iran, especially in the range to the north-east and north-west of Tehran. It inhabits rocky slopes at high altitudes, about 2300–4200 metres.

In cultivation subsp. *demavendica* is not a difficult plant although it is now becoming quite rare in gardens in spite of many introductions in the 1960s.

I am grateful to John Ingham for collecting some material of this plant in the mountains above Dizin in the Elburz mountains, and for making some temperature observations in its wild habitat. Where the plants were growing, at about 3200 metres, the soil temperature at midday on 5 and 26 October 1973, about 5cm below the surface, was 13°C (air temperatures were 14·5°C and 17°C on the respective days). On 26 October, there had been a frost at night and on 15 November, there was the first heavy snow on the site and this would have probably stayed until April or May the following year. Unfortunately he was unable to make a series of recordings over the whole period of one complete growth cycle, but these figures give an indication of how warm the soil stays, even well into the autumn, and the plants are then protected by snow for at least five months.

I. barnumae var. **zenobiae** Mouterde. This variety has now been divided by Shaukat Chaudhary into two species, *I. assadiana* and *I. swensoniana*. Descriptions of these can be found in the appropriate place in this section.

I. basaltica Dinsmore. A really vigorous species up to 70cm in height when in flower, it has rather tough leaves about 1·5–2cm wide and only slightly curved or erect. The very large flowers, 15cm or more in diameter, have falls which are heavily marked with blackish-purple veins and dots on a greenish-white ground, a rounded dark signal patch and a beard of maroon or brownish-purple hairs tipped with yellow. The standards are less heavily marked with lines and dots on a whitish or pale green ground. It is so named because it is endemic to stony basalt hillsides in Syria, where it flowers in April at an altitude of about 500m. The type locality is at the Krak de Chevaliers and it is known from a few other places, although Chaudhary notes that it is in danger of extinction.

It seems highly likely that the plant grown in Europe as *I. susiana* for several centuries is the same as either *I. sofarana* or *I. basaltica*, and represents one clone selected from wild populations. If this is so, the older name of *I. susiana* should take priority. However, much more work is required on this group before definite conclusions can be reached. It is sufficient to say that the description of the *I. susiana* of gardens would be so similar to that of *I. basaltica* given here that I have not repeated it under the former name.

I. **benjaminii** This is mentioned by Zvi Gazit-Ginsburg in the 1960 *Iris Year Book* as being very similar to *I. nazarena* but with less curled leaves and paler perianth segments with fewer dots. It is said to be from the Huleh region in the upper part of the Jordan valley. Since *I. nazarena* is in turn considered by both Mouterde and by Chaudhary to be the same as *I. bismarckiana*, it is obvious that there is still much confusion in this group of plants as to what constitutes a 'good' species and what is merely a colour form! I have given a description under *I. bismarckiana* and this will apply equally to the other names mentioned here.

I. **biggeri** Dinsmore. A robust plant up to 50cm in height with nearly straight leaves. The large flowers are described as having reddish-violet falls with dots which coalesce towards the base, and a darker signal patch also of a reddish-violet. The beard hairs are white with purple tips. The standards are paler, covered with fine purple veins on a white ground. It is recorded from rocky hillsides at Fakkuah in Palestine where it flowers in April.

I. **bismarckiana** E. Damman & C. Sprenger (Syn. *I. nazarena*). A strong-growing stoloniferous plant with stems up to 50cm in height, and broad, short leaves 2·3cm wide, spreading in a robust fan. The flowers, 10–12cm in diameter, have a creamy or yellowish ground colour densely covered with red-brown or purple spots and veins, and a large round blackish-purple signal patch on the falls. The beard consists of dark purple hairs. The standards are large and orbicular, white with many purple or blue dots and veins. It is a native of the southern slopes of Mount Hermon and nearby regions in north-eastern Israel.

I. **bostrensis** Mouterde. This is the same plant as Dinsmore's *I. atropurpurea* var. *purpurea*, raised to specific status by Mouterde and named after the town of Bosra. It is a smallish species for a southern Oncocyclus, with leaves less than 5mm wide and a stem about 10–15cm tall. The flowers are about 6–8cm in diameter, intensely spotted and veined deep purple-brown on a yellowish, greenish or pale brown ground. The signal patch is deep maroon and velvety and the dense beard is of yellow hairs sometimes tipped with purple. The species is a native of southern Syria in the regions bordering on Jordan. Chaudhary notes that it is the commonest of the 'black' irises of the region, often a cornfield weed. It flowers in April in the wild. Notes by R. W. Highwood of Beirut suggest that it is very variable in colour from pale reddish-brown with very faint minute dots to deep chocolate-brown. Some forms have golden yellow as a background, others are almost entirely uniform in colour and some are veined and finely dotted with the overall appearance of bronze.

I. camillae Grossheim. An extremely variable species in the wild, with many colour forms. It is 20–40cm in height with narrowly linear sickle-shaped leaves. The flowers are about 6–8cm in diameter with larger standards than falls. The colour can be pale yellow, bluish or violet, sometimes with a violet signal patch and with a yellow beard. There have been 15 colour forms described, some uniformly pale yellow, bright golden-yellow or pale blue while others are bicoloured with yellow falls and blue standards. In addition, the species mingles with *I. iberica* in the wild and many more variants have been described, probably of hybrid origin. *I. camillae* is a native of Transcaucasia, in the region of Lake Kazan Gel where it grows in stony places and flowers in April. It differs from *I. iberica* in its narrower, less rounded falls and in its yellow beard.

Some of the more interesting named forms include: *caerulea*—pale blue flowers; *lutea*—pale yellow flowers; *pallida*—white standards with falls densely veined; *speciosissima*—falls bronze, veined darker with a blackish signal patch and yellow beard; standards bluish; *spectabilis*—falls creamy with dense dark brown veining and a large blackish-brown signal patch; standards white; *sulphurea*—golden yellow falls with a reddish-violet signal patch; standards blue.

I. cedretii Dinsmore ex Chaudhary. This is usually 30–40cm in height and has straight or slightly curved leaves 1–2cm wide. The flowers are about 8–11cm in diameter and are white, densely but finely veined and spotted with maroon. On the falls is a rounded dark-maroon signal patch and a brownish or purplish beard of long hairs. It occurs only around the Cedars of Lebanon region at 2000 metres where it flowers in May.

I. damascena Mouterde. A rather compact plant for the size of the flower, it is usually only 15–30cm tall with curved leaves less than 1cm wide. The flowers are about 7–9cm in diameter, densely veined and spotted brown-purple on a whitish ground colour with a small dark purple signal patch. The beard consists of rather sparse purple hairs. The standards are usually paler than the falls because the veining is more purple than brown, and finer. There are many of these southern Oncocyclus with large veined and spotted flowers and they are difficult to distinguish from each other. *I. damascena* differs from the others (*I. sofarana* etc.) in having slightly smaller flowers in which the falls are rather long in proportion to their width, and in the shorter style branches (less than 5cm long). It is native to Syria, growing at about 1200m on the Jabl Qasyoun to the north of Damascus where it flowers in March or April. S. Chaudhary notes that it is in danger of extinction.

I. demavendica (Bornm.) Dykes. A synonym of *I. barnumae* subsp. *demavendica*.

I. eggeri Dinsmore. A synonym of *I. atropurpurea* var. *eggeri*.

I. elegantissima Sosnowsky. A synonym of *I. iberica* subsp. *elegantissima*.

I. elizabethae Siehe. Unfortunately, in the original description of this species, Siehe gave no indication as to the origin of his plant. One assumes however that it was from somewhere in southern or central Turkey since most of his collections were made there. It is described as being a dwarf plant with exceptionally long stolons. The leaves are sickle-shaped, about as long as the flower stem. The flowers have clear, straw-coloured falls with many broad purple-brown veins, a large deep purple-red signal patch and a bright yellow beard. They are said to be only slightly deflexed from the horizontal. The standards are similar in colour, a little broader than the falls, and acute, while the style branches are yellow-brown finely spotted and veined with red-brown. Siehe comments that this is closely related to his *I. sprengeri* (see below) and also resembles *I. acutiloba*.

I have collected a plant from the Nevşehir area of central Turkey, growing in the loose pumice dust slopes of extinct volcanoes. This has extremely long stolons, forming patches up to a metre across. It has flowered in cultivation and closely resembles *I. acutiloba* subsp. *lineolata* from northern Iran, which I have gathered on the pumice slopes of Kuh-i-Savalon. The habitats are very similar, and the Iranian plants also tended to be stoloniferous, although perhaps not as dramatically as the Turkish collection. It seems likely that the Nevşehir plant is *I. elizabethae* which in turn might well be synonymous with both *I. sprengeri* and *I. acutiloba* subsp. *lineolata*. The yellow beard of Siehe's species would set them aside from the latter of course and it seems preferable to retain the name *I. elizabethae* for the moment for the plant with long stolons from central Turkey.

I. ewbankiana Foster. A synonym of *I. acutiloba* subsp. *lineolata*.

I. fibrosa Freyn. A synonym of *I. meda*.

I. fominii Woronow. A synonym of *I. acutiloba*.

I. gatesii Foster. This is unusual in being a very large-flowered species but from a more northern area in Turkey and northern Iraq. The plant is extremely variable in colour in the wild, but always has nearly straight narrow leaves only 5–9mm wide and robust stems about 45–60cm in height. The enormous flowers can be 13–20cm in diameter and the overall impression of colour is greyish, brown or purplish depending upon the amount and colour of the fine stippling and veining on the segments. The beard is a rather diffuse band of hairs which may be yellow, brownish or sometimes purplish-brown and there is either no signal patch or only a very small insignificant dark one. In really pale forms the veins and dots are sparse and confined to the central portion of the falls and standards. In one collection I have seen, made by Oleg Polunin near Amadia in northern Iraq, the veining and spotting is very dense and the whole flower looks slightly bicoloured with brownish-veined falls and purplish-veined standards.

I am indebted to John Watson and Adil Güner for the loan of colour transparencies showing variations in a population of *I. gatesii* from Halkis Dag. These showed that the ground colour varies from almost pure white to cream, sometimes tinged with pink. The markings on the falls basically consist of brown, reddish or blackish spotting and veining, sometimes the spots suppressed with only the veining remaining and sometimes with hardly any veining but with much spotting. The signal patch varies from a tiny narrow pale brown or blackish mark right in the centre of the falls to a much larger band or roundish patch, but always small in relation to the large size of the falls. In one specimen there is a pale violet 'halo' around a blackish circular patch. The beard varies from yellow to dark brown, usually sparse, but in one specimen it is confined to a more dense and clearly defined band. The standards on the whole follow the same colour as the falls, but tend to be more veined than spotted and the veins coalesce in the centre and on the haft to form a darker stain. One photograph showed an impressive clump of this enormous iris with ten flowers.

I. gatesii occurs in south-eastern Turkey, and north-eastern Iraq. It grows in rocky limestone hills at 1100–2000 metres and flowers surprisingly late, in May or June in the wild.

I. grossheimii Woronow. A Caucasian species which I have not seen, other than as a herbarium specimen which shows a dwarf plant about 13cm in height with a flower somewhat larger than that of *I. acutiloba*, about 7–8cm wide. Dr Rodionenko describes it as having the standards much larger than the falls. The colour is wine-red or dark brown with cinnamon-purple veins, and on the falls, which are only 5cm long and 2cm wide, there is a large blackish-brown signal patch. It is apparently a beautiful iris which inhabits hills and mountain slopes in the Nakhichevan region of southern Transcaucasia.

I. hauranensis Dinsmore. This is considered by Mouterde to be a synonym of *I. jordana*. It was described from the banks of the river Jordan, in fact from the opposite bank to *I. jordana* and the type localities of the two 'species' are very close to each other. Dinsmore regarded *I. hauranensis* as having rather more purple-coloured flowers with a less intense signal spot and a beard with short white purple-tipped hairs, whereas the beard of *I. jordana* is of longer yellow hairs. These features probably merely represent variation within the species from population to population.

I. haynei (Baker) Mallet. This is another of the large-flowered southern Oncocyclus which have flowers 10–12cm in diameter, veined and spotted on a pale ground colour. In this case there is much less spotting but the segments are beautifully and densely brown-veined. The signal patch is intensely blackish-brown and the standards are more purplish in colour. Dinsmore in

Post's Flora of Syria, Palestine and Sinai considers it is a doubtful species although it seems to be maintained by some of the present day botanists in Israel. It is endemic to Mount Gilboa, north-east of Jinin. where it flowers in April at an altitude of 150–350 metres in rocky limestone scrub. P. H. Davis describes the overall appearance of the flowers as dusky lilac and notes that they were fragrant.

I. helena (Koch) Koch. A synonym of *I. acutiloba* subsp. *lineolata*.

I. helenae Barbey. A synonym of *I. maria* Barbey.

I. hermona Dinsmore. Unlike the similar *I. bismarckiana* this is a non-stoloniferous plant. In overall appearance it is also like *I. sofarana*. The leaves are more than 1cm wide and rather straight and erect. The flowers are about 10cm in diameter and have creamy yellow falls with brownish-purple veining, although this is rather less dense than in related species so that the cream or yellowish colour predominates. The rounded signal patch is purplish-brown and the beard hairs are rather sparse and brown-purple. Normally the standards are cream or whitish with fewer veins and dots of purple. As its name suggests the species occurs on Mount Hermon and adjacent areas and is April-flowering.

I. heylandiana Boiss. & Reut. ex Boiss. This name was originally used to describe a plant said to grow in south-east Turkey and northern Iraq. However it has since become clear that the plants from these two areas are different and that the name actually covers two species. Under the rules of nomenclature it is possible to select one of the specimens as the type material of the name *I. heylandiana* and S. A. Chaudhary did this in 1977, selecting the specimen from Iraq (between Mosul and Baghdad). The Turkish specimen is the same species as that described later by Foster as *I. gatesii* and can therefore take this name.

   I. heylandiana consequently applies to the Oncocyclus iris growing in Iraq, south of Mosul, but its distribution and characters are poorly known, owing to lack of good collections in the area. From the type specimens the following observations can be made. It is about 40cm in height with rather short curved leaves about 1–1·2cm wide which sheath the flower stem. The flower is about 8–9cm in diameter and appears to be veined on a pale ground, possibly with an overall pale brownish appearance. The beard consists of rather long straggling hairs. Unfortunately it is not known if the rhizome is compact or stoloniferous, for this would be of interest in comparing it with other species in the region.

   I. nectarifera, a recently described plant from the Turkish–Syrian border area, is obviously closely allied to *I. heylandiana* but can be distinguished by its narrower falls. It is a very strongly stoloniferous plant and apparently also occurs in adjacent Syria and north-western Iraq around Balad Sinjar and Tel Afar. A further description will be found under *I. nectarifera*.

   I. heylandiana is therefore still a little-known plant and new collections are badly needed on the Mosul–Baghdad road in order to resolve the problem. Occurring in this region, it follows that it must be a plant of almost semi-desert conditions, or at least low hills which are sun-baked in summer.

I. iberica Hoffm. A rather dramatic plant from the Caucasus, unfortunately not an easy plant to maintain in cultivation. It grows 15–20cm in height and has narrow, strongly curved grey-green leaves. The flowers are bicoloured, the falls being heavily brown-veined and spotted on a pale ground with a round, blackish or intense-brown signal patch and a beard of long, purple-brown hairs. The standards are white, creamy or pale bluish and much less veined. The type locality of *I. iberica* is Tiflis, and the particular variant known as subsp. *iberica* is confined to Russia. Further to the south it is replaced by two other subspecies as follows.

I. iberica subsp. *elegantissima* (Sosn.) Fedor. & Takht. This is generally larger-flowered than subsp. *iberica*, the flowers up to 10cm in diameter, and the stems are usually 20–30cm in height. The falls reflex very sharply so that the blade lies almost vertically. In subsp. *iberica* the falls are less vertical and are usually deflexed at an angle of about 60° from the horizontal. Although the overall colouring is similar to that of subsp. *iberica*, the ground is usually creamy or even pale yellowish and I have not seen any forms with bluish-tinged standards. In some forms there is considerable brown-veining on the standards also, although seldom are they similar to the falls in the amount and density of the markings, so the bi-coloured effect is maintained. However, one form which I grow, collected by my colleague Wessel Marais near Tortum in Turkey, has strongly veined standards. Subsp. *elegantissima* was described from the Kars region, which is in north-east Turkey, and is very common as far west as Erzurum and south to Lake Van. It is also in north-western Iran and adjacent Russian Armenia. It is a very striking plant, an amazing sight when in flower on the open rocky slopes of such mountainsides as that of Ararat (Ağri Dağ). It occurs at altitudes of about 1000–2000 metres and flowers in April.

I. iberica subsp. *lycotis* (Woron.) Takht. This is a widespread variant, differing in having the falls and standards more or less equally heavily veined and dotted so that the whole flower often appears brownish or purplish. The falls stand out at an angle rather than being deflexed to the vertical and are very markedly concave with a velvety brown signal patch in the centre. It occurs in Armenia, north-west and western Iran as far south as Isfahan, north-eastern Iraq and south-eastern Turkey. In some forms the veining and spotting is almost pinkish, while one collection from west Iran is described as having rose-coloured flowers with a burnt orange hood (presumably referring to the style branches). Subsp. *lycotis* occurs in fields and on stony hillsides at 1450–3000 metres and flowers in April or May.

Paul Furse introduced this subspecies from many different regions in the early 1960s, but it is still a rare plant in cultivation.

I. iberica var. *bellii* Baker is apparently the same as subsp. *lycotis*.

The Russian floras describe great variety in subsp. *lycotis* (which most of their botanists have treated as a separate species from *I. iberica*). For example there is var. *heterochroa* which has falls with a ground colour of brownish-cherry and lighter standards; var. *magnifica* with a dark cherry-red ground colour; var. *panthera* with prominent spots on a lighter ground and var. *pardus* with the standards veined and the falls spotted. The most common form with veined falls and standards was called var. *typica*, but under present-day rules this should be called *I. iberica* subsp. *lycotis* var. *lycotis*.

I. jordana Dinsmore. This is one of the few or perhaps the only Oncocyclus iris which can claim to grow below sea level! It is recorded from −250 metres in the Jordan river valley to the south and south-east of the Sea of Galilee. It is a robust species up to 45cm in height with rather erect leaves which sheath the stem completely. The flower is about 12–15cm in diameter and appears purplish-maroon or sometimes pinkish because of the density of the lines and spots on the white ground. The signal patch is rounded and velvety blackish-purple and the beard is of long creamy-yellow hairs. It grows on plains and flowers in April. *I. hauranensis*, from the plateau of Hauran to the east of Lake Tiberias (Sea of Galilee), is considered by Mouterde to be a variant of this species.

I. kasruwana Dinsmore. A synonym of *I. sofarana* subsp. *kasruwana*.

I. kazachensis Grossheim. This is regarded by Dr Rodionenko as a hybrid between *I. acutiloba* subsp. *lineolata* and *I. paradoxa*, both of which occur in Transcaucasia, Iran and Turkey. Wherever Oncocyclus iris species meet in the wild there are likely to be such hybrids.

I. keredjensis Parsa. This is a distinct-looking Oncocyclus but I know it only from the original herbarium specimen which shows some affinities with *I. barnumae*. It has very erect, narrow leaves and is about 30cm in height when in flower. The flowers are about 8–9cm in diameter and look as if they could have been lilac or bluish when fresh, although Parsa describes them as being white. The dense beard is deep violet and in this respect it resembles *I. barnumae* forma *protonyma*. It was collected in Iran on 20 May 1941 by Parsa, at 2200 metres in the Elburz mountains, near Karadj.

I. kirkwoodii Chaudhary. One of the large-flowered heavily spotted and veined irises from Syria. This is a very tall one, sometimes as much as 75cm in height with curved, rather weak, often drooping leaves, 1–1·5cm wide. The flowers are about 8–12cm in diameter and are white or pale greenish covered with dark purple veins and dots. The signal patch is rounded and deep purple, and the beard consists of long purple or brownish-purple hairs. In the typical form the standards are more of a pale blue shade with darker veins and speckling. *I. kirkwoodii* grows in the area of Bishmishly in northern Syria, and in the adjacent Amanus mountains of Turkey where it has been collected by S. Albury, M. Cheese and J. Watson and much earlier by E. K. Balls at Maraş. Plants from the former expedition in 1966 may still be in cultivation in Britain and the United States under the number A.C.W. 845. These were growing at about 750 metres on limestone hills. The standards were densely veined violet on a white ground and the falls were stippled purple on a cream base, with a purplish-black beard. They were in flower in the wild at the end of April.

I. kirkwoodii var. **macrotepala** Chaudhary. This is a variant from a little farther to the south around El Bara, although still in northern Syria. The falls are larger, about 10cm long (6–8cm in var. *kirkwoodii*). The beard can be either purple or golden.

I. kirkwoodii subsp. *calcarea* Dinsmore ex Chaudhary. Another variant from northern Syria at Deir Semaan between Aleppo and the Turkish border. The flowers are about 8cm in diameter and the falls are said to be more obovate in shape than ovate. The colour is described as being dotted and veined dark purplish-red on a pale green ground with a velvety maroon signal patch and a sparse beard of long maroon hairs. The standards are light blue, veined with darker purple. This variant seems to grow reasonably well in Surrey, although not flowering every year.

I. koenigii Sosnowsky. A hybrid from the Caucasus, throught to be *I. paradoxa* × *I. iberica*.

I. lineolata (Trautv.) Grossheim. A synonym of *I. acutiloba* subsp. *lineolata*.

I. lortetii Barbey. A rather dramatic species, probably now very rare in the wild. It is about 30–50cm tall when in flower and has rather straight leaves about 1–1·5cm wide, almost parallel-sided until the abruptly narrowed tip. The flower which is about 8–9cm wide, is usually pinkish in appearance because of the fine pink or maroon spotting on a white ground. The standards are similarly coloured but delicately veined rather than spotted. The signal patch is a striking deep maroon and the beard hairs are rather sparse and reddish. This attractive plant is native to southern Lebanon in the border area with Israel, flowering in May, and it grows in rocky places, sometimes in scrub. Fortunately it is not excessively rare in cultivation and is occasionally available in the nursery trade.

I. lupina Foster. This names applies to a plant from Harput near Elaziğ in central Turkey. It is a rather vigorous variant of *I. sari*, a very widespread species in Turkey, and a description will be found under that name. It is recorded by Foster that the local name in Turkey is the 'Wolf's Ear', referring apparently to the tawny-coloured tips of the standards in some of the forms, hence the epithet 'lupina' which was given by him.

I. lycotis Woronow. A synonym of *I. iberica* subsp. *lycotis*.

I. maculata Baker. This is a name best ignored, for the description is poor and the original specimen is so badly damaged that one can say no more than that it is an Oncocyclus iris with the vague locality of 'Mesopotamia'. It is possibly the same plant as that described by Boissier as *I. heylandiana* or the recently described *I. nectarifera*, but to equate it with either of these would be guesswork.

I. manissadjianii Freyn. This appears to be synonymous with *I. sari*. It is a small form from Amasya in Turkey, only 10–15cm in height.

I. mariae Barbey (Syn. *I. helenae* Barbey). One of the most southerly-occurring species of Oncocyclus. It produces small rhizomes which possess a lot of fibrous remains of old leaf bases, and is about 15–25cm in height when in flower. The leaves are only 3–6mm wide and are strongly arched. The flowers are 8–10cm in diameter and are lilac or pinkish with a deeper purple region on the haft of the falls which have a blackish-violet signal patch and a deep purple beard. The standards are much larger than the falls and are lilac-purple to pinkish or violet with deeper veining. It inhabits southern Israel on the Sinai peninsular and is recorded as far south as the Tih plateau. It grows in sandy places and flowers in March.

I. meda Stapf. (Syn. *I. fibrosa*). A beautiful and distinctive Persian endemic, aptly described by Paul Furse as 'Honey Gold' in some of his papers. It is a slender species, about 10–25cm in height with strongly curved to nearly straight leaves only 1·5–4mm wide. The flower is about 5–7cm in diameter and varies a lot in colour, but the species is usually recognizable because of the irregularly undulate segments. The ground colour is cream or whitish but this is largely masked by golden brown veining and suffusion, especially towards the edge of the segments. The signal patch is large and darker brown or purplish, and the beard is a very prominent one of long yellow hairs. Occasionally there is a lilac tinge to the flower and sometimes the overall appearance is yellowish-brown. *I. meda* flowers in April and May on stony hillsides and at the edges of fields in western Iran at altitudes of 1300–2200 metres. It is particularly common from Mianeh and Karadj south-west to Hamadan and Sanandaj where it appears to meet and hybridize with *I. barnumae* forma *protonyma* giving a whole range of curious plants. This area, however, needs a thorough investigation before one can be sure about this statement. Typical *I. meda* is a delightful plant with its warm brown colours and crinkled perianth segments, and does not seem to be too difficult to grow. I have a pleasing form collected by John Ingham which has proved to be fairly amenable to cultivation.

I. medwedewii Fomin. A synonym of *I. paradoxa* forma *choschab*.

I. nazarena (Foster ex Herb.) Dinsmore. A synonym of *I. bismarckiana*.

I. nectarifera Güner. A recently described species from the borders of Turkey and Syria, it is a vigorous plant about 25–40cm tall in flower and produces rhizomes with very long thick stolons. The leaves are curved and about 1–1·3cm wide in strong specimens. The flower is about 8–9cm in diameter and is creamy-white, or yellowish, delicately but prominently veined with brown or purplish-brown. The signal patch is brownish-maroon and the beard is a narrow but dense yellow strip of hairs. It occurs to the south of Mardin around Şenyurt on the Syrian border where it grows at about 500 metres in steppe country and flowers in April. The species is distinguished by Dr Güner on its stoloniferous habit and in having nectaries on either side of the base of each of the falls. Var. *mardinensis* is a variety from the same locality, separated from var. *nectarifera* in having a shorter perianth tube (2–2·5cm versus 3·5–4cm) and narrower basal leaves (0·8–0·9cm versus 0·9–1·3cm).

I am indebted to Adil Güner from Ankara for sending specimens, photographs and a rhizome of this species. As yet it has not flowered in cultivation.

It is undoubtedly close to *I. heylandiana* which is a rather poorly known plant from south of Mosul in Iraq, but this species apparently has wider falls (3·5cm) than in *I. nectarifera* (2·3–2·5cm). *I. nectarifera* appears to grow on the Syrian side of the border at Derbassieh and probably also in Iraq, for I have seen specimens from the Sinjar and Tel Afar regions which resemble it closely. The Derbassieh material has been cultivated in America for some years and photographs I have seen which were taken by Mr Clay Osborne and Mr Herbert McKusick show an iris very similar to that from the Turkish side of the border, and it is said to be long-stoloniferous.

I. **nigricans** Dinsmore. Another one of the so-called 'black' irises of Jordan, this produces a smallish rhizome, grows to about 20–30cm in height and has narrow, very recurved leaves. The flower is about 8–10cm in diameter and is dark brownish-purple, or sometimes blackish-purple, with darker veins and dots. It has a black signal patch on the falls and the beard hairs are dark-purple tipped, on a whitish ground colour. The standards are uniformly covered with dark purple-lilac veins on a white background although usually very little of this shows through. It flowers in April in the wild and grows in cornfields and on the edges of cultivated places at about 750–900 metres altitude. It is mainly in the area of Jordan to the east of the Dead Sea, south of Amman.

I have not succeeded in its cultivation and other writers report it to be difficult to keep.

I. **paradoxa** Steven. This is probably one of the most distinctive of all the Oncocyclus irises and is immediately recognizable in any of its forms. It grows to about 10–25cm in height and has strongly curved leaves only 2–4mm wide. The flower is remarkable in that the nearly horizontal falls are very small compared to the large rounded standards, which are 8–10cm long, and the blade of the falls is almost entirely covered with a dense beard. The colour varies considerably and several names exist for these variants. In Armenia there is a common form with purple or violet standards and darker falls with a blackish beard, and one (forma *atrata*) which has blackish-violet standards as well as the falls. One of the most beautiful is forma *mirabilis* in which the golden yellow falls are covered with a darker, almost orange beard and the standards are paler yellow or pale blue. Paul Furse collected one similar to this in Iran near the border town of Julfa. In eastern Turkey, north-western Iran and adjacent USSR there is forma *choschab* in which the falls are blackish-violet and the standards white with delicate dark veining. Normally these all have a dark signal patch, but occasionally the yellow forms lack it altogether. No doubt a thorough search of the whole area of Transcaucasia, northern Iran and eastern Turkey would reveal more forms. In the wild it normally flowers in May or June and inhabits stony hillsides and banks near fields at 1200–2000 metres. I have seen it in great quantity on the hills to the south-east of Lake Van, always in the *choschab* form, and in fact I have not seen a record of any other form from Turkey, although Grey in his *Hardy Bulbs* mentioned a form with white unveined standards.

In cultivation, *I. paradoxa* is one of the easiest species to grow and makes a beautiful alpine house plant or bulb frame subject.

I. petrana Dinsmore. A dark-flowered species from Jordan resembling *I. nigricans*, but the narrow leaves are only slightly curved. The flower is smaller, only about 7–8cm in diameter and is deep lilac with little veining but with a blackish signal patch and a dense beard of purple-tipped hairs on a yellow ground. The standards are also dark lilac but the style branches are a rather more brownish-lilac shade. *I. petrana* flowers in April and grows on semi-desert plains and limestone slopes in Jordan to the east and south of the Dead Sea, at about 1200–1500 metres.

A colleague, Dr Daoud El-Eisawi, has observed and collected these dark-flowered irises of Jordan and says that the range of variation is such that it seems unlikely that there are several species involved.

I. polakii Stapf. This is a mystery case, for in spite of a search of various herbaria, the original specimen has not been located. From the description alone it is not possible to identify the plant with certainty although the name could refer to one of the many curious Oncocyclus irises which occur around Hamadan and Sanandaj in western Iran, possibly hybrids between *I. meda* and *I. barnumae* (mentioned above under *I. barnumae* forma *protonyma*). The name *I. polakii* has been much used in literature when referring to the latter plant, although there is not real justification for this.

I. polakii var. **protonyma** Stapf. A synonym of *I. barnumae* forma *protonyma*.

I. samariae Dinsmore. A robust species having a stout rhizome with many bristle-like fibres at the apex and flower stems usually about 25–30cm tall. The leaves clothe the stem throughout and are nearly straight and less than 1cm wide. The flowers are about 9–10cm in diameter in the specimens I have seen and they are purple-veined and spotted on a creamy white ground with the standards usually a pinkish-purple. The beard is of sparse yellow-brown hairs which become shorter and purple at the edge of the beard. The signal patch is small and dark brown. It is very similar to *I. lortetii* but the falls are not as strongly reflexed and the markings consist of much larger spots. *I. samariae* occurs on rocky limestone slopes around Nablus at about 800 metres and it flowers in April.

I. sari Schott ex Baker. Although J. G. Baker when describing this species stated that it had lilac flowers, he was apparently using a dried specimen and it is well known that brownish-purple colours in Oncocyclus iris can change to a bright lilac on drying. From his specimen at Kew there seems no reason to doubt that his plant is the same as the brownish-purple iris which occurs widely in Turkey. Baker's specimen came to him from the nurseryman Max Leichtlin and the only information given as to its wild origin was 'Persia or Armenia'.

I. sari is one of the easiest and more robust of the Turkish species, growing 10–30cm in height with nearly straight or curved leaves. The flowers are about 7–10cm in diameter and are usually veined crimson or brownish-purple on a yellowish, creamy or greenish ground. The signal patch is an intense crimson or brown stain and the beard is always yellow. Usually the standards are the same colour as or darker than the falls, but sometimes they are bluish-purple. Both falls and standards are often undulate, giving the flower a ruffled appearance. It occurs in central and eastern Turkey from the Ankara area to Lake Van, growing on stony hills between 1000 and 2250 metres and flowering from April to June.

I. sari varies enormously in colour and stature, often within one population, from dwarf plants with very curved leaves to robust ones with nearly straight leaves, depending upon the habitat of the individual plants. The plants from around Elazığ, Erzincan and Erzurum are particularly robust and represent the form which was described as *I. lupina* by Foster.

In cultivation *I. sari* is a fairly easy species to grow and keep, although it requires bulb frame protection in Britain to give it the summer dormant period it requires.

I. schelkownikowii Fomin. This is a relative of *I. acutiloba* with pointed tips to its perianth segments. It is, like that species, a dwarf rather small-flowered plant and is known in several colour forms. The flowers are a little larger than those of *I. acutiloba* and usually have a brownish or violet overall appearance with slender darker veins. The standards are usually darker and much larger than the falls. The signal patch is purple and the beard is yellow. The flowers have, according to Rodionenko, a delicate scent. It is a native of Russian Azerbaijan where it grows on the banks of the river Kura in low hills on clay soils.

I. sinistra Sosnowsky. A plant occurring in Transcaucasia, thought to be an occasional hybrid of a non-persistent nature.

I. sofarana Foster. This is one of the best known in the group of southern Oncocyclus species which have heavily veined flowers. It grows about 30–40cm in height and has strong wide leaves up to 2·5cm broad, usually straight or nearly so. The flowers are 10–13cm in diameter and are densely purple-brown or violet veined and spotted on a creamy ground. The rounded signal patch is blackish-purple and the beard is rather sparse and also dark purple. Normally the standards are paler than the falls with a white background and finer violet or reddish-purple veins and speckling. The typical form occurs in Lebanon on the mountains between Beirut and Damascus, flowering in May. It is probably now very rare in the wild.

I. sofarana forma **franjieh** Chaudhary. A variant occuring in populations of the typical form with pale flowers with varying amounts of yellow on the falls and standards, which are otherwise white. There is none of the violet or purple colouring present in the flower.

I. sofarana subsp. **kasruwana** (Dinsmore) Chaudhary. This is very similar in its colouring to typical *I. sofarana* but the standards are usually nearly as heavily veined as the falls. It is separated by Chaudhary mainly on the combined width of the two style lobes which is about as wide as the lower curved part of the style branch. In the other variants of *I. sofarana* the combined width is more than that of the style branch. The signal patch of subsp. *kasruwana* is much narrower (0·6–1·5cm) and is tear-drop shaped, whereas in subsp. *sofarana* it is nearly round.

Subsp. *kasruwana* is known from only two places in Lebanon, Naba-al-Asal and Laqlouq. It grows on limestone rocks at about 1750 metres and flowers in May. A collection by Dr Chaudhary (number 790) flowered in 1980 for the first time in my cold greenhouse. Its flowers were 10cm in diameter and the veining was of a dull purple. It is a very strong-growing plant requiring much feeding in the growing season.

I. sofarana forma **westii** (Dinsm.) Scaly. A synonym of *I. westii*.

I. sprengeri Siehe. This was found in 1903 by Siehe and described by him, with a photograph, in the *Gardener's Chronicle* of 1904. It is a small plant, only 10cm in height with creeping rhizomes forming stolons. The falls of the flower are yellow with bright purple-red spots and veins and with a golden-yellow beard. The standards are silvery-white-veined with purple-red and black. The photograph in the *Gardener's Chronicle* shows a little plant rather similar in appearance to *I. acutiloba* subsp. *lineolata*, but the yellow beard would separate it from this. With the type of creeping rhizomes which Siehe describes, it is very difficult to find any difference between *I. sprengeri* and *I. elizabethae*, also described by him.

I. strausii Micheli. An Iranian plant, it was described in 1899 from material collected at Arak and cultivated by Leichtlin. Unfortunately the description is too vague to give any idea of its identity.

I. susiana Linn. The Mourning Iris. This is horticulturally the most well known of the Oncocyclus irises for it has been in cultivation for at least 400 years. It is reported that it was sent by Busbecq, the Austrian Ambassador in Turkey, to Vienna in 1573 and from there reached other places in Europe. Almost certainly it has been vegetatively propagated ever since and the plant we have today is probably the same clone as that introduced from Constantinople so long ago. In characters it differs from the Turkish wild species and it is likely that the plant was being cultivated by the very garden-conscious Ottoman Turks. Dr West of Beirut, writing in the *Iris Year Book* of 1953, pointed out that 'sūsan' is an Arabic word for 'iris' so the name is almost certainly derived from that. Furthermore the *I. susiana* of commerce bears a great resemblance to *I. sofarana* and *I. basaltica* and may be a form of one of these which was selected and grown by the Turks. It is known that the Oncocyclus irises of the Lebanon and adjacent countries are now very rare and some of them very restricted in their range and near extinction. It is quite probable therefore that the population from which our *I. susiana* was originally selected is not now in existence.

I. susiana has a large flower heavily veined with deep purple on a greyish ground, with a velvety black signal patch and deep purple beard. The leaves are only slightly curved. Unfortunately most, if not all, the stocks of this plant are now infected with virus.

I. swensoniana Chaudhary, Kirkwood and Weymouth. A recently described species from Syria. It is about 40cm in height with very strongly curved leaves less than 1cm wide. The plant is clump-forming but not stoloniferous and the rhizomes are quite small. The flowers are about 7–8cm in diameter and have almost blackish-purple falls and slightly lighter purple standards. The signal patch is an even darker velvety blackish-maroon and the beard is of contrasting yellow hairs tipped with purple. This very dark overall appearance is broken by the style branches which are orange streaked with purple. Obviously this is a striking species one would like to see in the wild but it is endemic to the Tell Chehan area in southern Syria and this is probably not the easiest of places in which to travel at the time of writing.

I. szovitsii C. A. Mey. A synonym of *I. acutiloba*.

I. tatianae Grossheim. A name given, according to Rodionenko, to a non-persistent hybrid occurring in the Caucasus.

I. urmiensis Hoog. A synonym of *I. barnumae* forma *urmiensis*.

I. westii Dinsmore. A plant with a very mixed taxonomic career, having been treated as a species or, at the other end of the scale, as a mere colour form of *I. sofarana*. Not having studied the wild populations of Lebanese irises I can only accept the most recent work in which it is restored to species status. It grows about 30cm in height and has slightly curved leaves less than 1cm wide. The colourful flowers are about 12–15cm in diameter and have the falls heavily chocolate- or purplish-blotched and veined on a pale yellowish ground colour, and contrasting standards in pale lilac with darker lilac-blue veins and dots. The signal patch is a deep velvety chocolate and the purple beard hairs are rather long and sparsely arranged. It is a beautiful plant, unfortunately now rather rare in cultivation. At one time it was grown very well by Eliot Hodgkin and a plant of his was used for the fine illustration in the *Botanical Magazine*, tab. 550 (1969). The markings on the falls consist of rather large blotches, closely spaced so that the yellowish ground appears through as a reticulated pattern. In many of the irises of this sort the markings are finer, appearing as a dense speckling on a paler ground.

   I. westii occurs on the mountains between Jezzine and Mashghara in the Lebanon.

I. yebrudii Dinsmore ex Chaudhary. A Syrian species of short stature usually 15–18cm in height with rather short, stiff, slightly curved leaves which are said to be noticeable for their grey-white 'bloom' on the surface. The flower is about 8–9cm in diameter and has the falls veined and speckled brown-purple on a yellow ground. The signal patch is quite small and dark purple, and the beard is of long purple hairs. The standards are yellowish, less prominently veined purple. It grows in the Yebrud region of Syria, flowering in May.

I. yebrudii subsp. *edgecombii* Chaudhary. This is a variant with larger flowers, about 10–12cm in diameter, with the falls covered with red-purple dots and broken lines on a pale yellow or greenish ground, a maroon-purple signal patch and a beard of purple, yellow-tipped hairs. The standards are more densely veined maroon on a clear white ground. It is endemic to the Kastel region of Syria and is noted by Chaudhary to be in danger of extinction. Some years ago Dr Chaudhary sent me a rhizome from his cultivated stock of this and it has proved to be a relatively easy species to grow although not, it seems, very hardy. I lost most of the rhizomes in a frosty period one winter. Like most of the irises of this type it is quite a striking plant, and not too tall.

I. zuvandicus Grossheim. Another of the Caucasian plants which Rodionenko regards as being a sporadic natural hybrid.

1D Subgenus Iris, Section Regelia

A relatively small group of species, the Regelia irises are closely related to the Oncocyclus but distinguished from them by having (nearly always) two flowers on each stem rather than solitary ones, and a beard on all six of the perianth segments (falls and standards) instead of just the three falls. They are also apparently distinct in their chromosome structure, and are separated geographically. Travelling eastwards, the Regelias first appear in north-east Iran where the Oncocyclus end, and they continue through northern Afghanistan and adjacent Russian Central Asia—a comparatively small overall area of distribution for the eight or so species.

The Regelias are also allied to the *Psammiris, Hexapogon,* and *Pseudoregelia* Sections and more information can be found about the relationships and differences under these headings.

Cultivation

Like the Oncocyclus, Regelia irises occur on open rocky or sandy hillsides where they have a long summer dormant period. However, several of them seem to be able to tolerate the British weather rather better than Oncocyclus and a few such as *I. korolkowii* and *I. hoogiana* can be grown outside in well-drained sunny places. The best method of cultivation however is in a bulb frame as described for the Oncocyclus group (see page 42) and I will not repeat it here. For the Regelia hybrids, some of which are crossed with pogon species, it is better to grow them in an outside bed as they do not appear to like a summer baking. On the whole the Regelia species are not too difficult in much of Britain and the United States but there are unfortunately few which are readily obtainable.

The Regelia iris species

I. afghanica Wendelbo. This delightful species was described as recently as 1972 and is in cultivation although rather rare, and unfortunately some stocks are already infected with virus. It is about 15–35cm in height, forming small tufts, with somewhat curved leaves. This is the one species of Regelia which often produces a solitary flower although two are occasionally present. In all other respects however it is a 'good' Regelia. The flower is about 8–9cm in diameter and is bicoloured. The rather pointed falls have a creamy or white ground colour heavily veined purple-brown with a solid purple signal patch in the centre and a beard of long dark hairs. The standards are also fairly pointed and are usually pale yellow with a beard of greenish hairs on the lower part. It occurs in north-eastern Afghanistan, particularly on the Salang Pass but also in other places in Kataghan province, at altitudes of 1500–3300 metres. Its habitat is among rocks on mountain slopes, apparently always on granite and shale formations, where it flowers in May and June.

It seems to be a relatively easy plant to grow in an alpine house or bulb frame.

I. darwasica Regel (Syn. *I. suworowii*). A little-known plant, it is, together with *I. lineata* which it resembles, in need of further study in the wild. It is 15–40cm in height, tufted in habit with the fibrous remains of old leaves attached to the shortly stoloniferous rhizomes. The leaves are straight and about 4–8mm wide. The flowers are about 5–6cm in diameter and are said to be purplish-veined on a lilac ground colour with a beard of purple hairs on the falls. It seems to be separated from *I. lineata* by its longer falls and standards (6–7cm long), which are rounded at the apex (acute or subacute in *I. lineata*), and in the slightly broader leaves. It is a native of Russian Central Asia around Bokhara and Darwas, close to the border with Afghanistan, and is a plant of rocky mountain slopes, flowering in June.

I. darwasica has been in cultivation but I cannot find any recent reference to it in Britain. It is quite probable that at least some of the plants grown in the past under this name were in fact *I. lineata*. Dykes comments on the 'long pointed segments' of his *I. darwasica* which indicates that he was in all probability growing *I. lineata*, although the two species are so poorly known that they may just be variants of the one species.

I. heweri Grey-Wilson & Mathew. A delightful dwarf Regelia collected and introduced to cultivation by Christopher Grey-Wilson and Tom Hewer (nos. 746 & 757 and 769). It is only about 10–15cm, rarely 30cm, in height and has thin rhizomes producing stolons up to 5cm long. The leaves are strongly curved and only 2–5mm wide. The one or two flowers per spathe are about 5cm in diameter with deep purple-blue falls which have a whitish, slightly purple-veined haft and a lilac beard. The standards are also deep or mid-purple with a paler beard at the base—in fact the flower is almost uniform in colour without prominent veining. It is only known from north-eastern Afghanistan in Kataghan province where it flowers in May on grassy mountain slopes, in scree or rather sandy hillsides at 1150–2200 metres.

I. heweri is a distinctive species, apparently not too difficult to grow in pots or bulb frames in Britain, although, as with many of the iris species from Afghanistan, it is still very rare in cultivation.

I. hoogiana Dykes. A robust species and one of the easier ones to grow in Britain. It grows 40–60cm in height with erect straight or slightly curved

1 Bulb frames at
Kew
2 Irises of
Subgenus Scorpiris
in pots

SUBGENUS IRIS, SECTION IRIS 3 *I. reichenbachii* near Titov Veles, Yugoslavia

SUBGENUS IRIS,
SECTION IRIS
4 *I. suaveolens*
5 *I. lutescens*

SUBGENUS IRIS,
SECTION IRIS
6 and 7 *I. schachtii*
from the Beynam
Forest, Ankara,
Turkey

SUBGENUS IRIS,
SECTION IRIS
8 and 9 *I. taochia*
near Tortum,
Turkey

SUBGENUS IRIS,
SECTION
ONCOCYCLUS
14 and 15 *I. iberica*
subsp. *elegantissima*
in Turkey

SUBGENUS IRIS,
SECTION
ONCOCYCLUS
16 and 17 *I. sari*

SUBGENUS IRIS,
SECTION
ONCOCYLUS
18 *I. gatesii*
19 *I. susiana*

SUBGENUS IRIS, SECTION REGELIA 20 *I. hoogiana*

SUBGENUS IRIS, SECTION LOPHIRIS
21 (above) *I. tectorum* 'Alba'

SUBGENUS LIMNIRIS, SERIES SPURIAE
22 (above right) *I. orientalis* near Izmit, Turkey
23 (right) *I. spuria* subsp. *musulmanica* near Agri, Turkey

SUBGENUS LIMNIRIS, SERIES SYRIACAE 24 (above left) *I. masia*; SUBGENUS SCORPIRIS 25 (above right) *I. magnifica* bulb, showing fleshy roots; 26 (below) *I. aucheri*

Subgenus Scorpiris 27 and 28 (above) *I. bucharica*
29 (below) *I. persica*

SUBGENUS
SCORPIRIS
30 (right)
I. caucasica

SUBGENUS
HERMODACTY-
LOIDES
31 (below left)
I. vartanii 'Alba';
32 (below right)
I. winowgradowii

purplish-tinged leaves 1–1·5cm wide. The rhizomes are thick and produce long, thinner, stolons. The spathes produce two or three large scented flowers which are 7–10cm in diameter and coloured a plain lilac-blue with a yellow beard. A white form is known but whether or not it is still in cultivation I do not know.

The original collection of *I. hoogiana* was made in 1913 by Graeber who was collecting for the firm of Van Tubergen, and the only data given was that it came from Turkestan. However it is now known that it occurs wild on the slopes of the Pamir-Alai mountains and has been gathered in the Varzob valley, flowering in June.

It is a fine and graceful species with clear-coloured flowers of a delicate texture. There are some named selections differing slightly in colour, and some hybrids with *I. stolonifera*.

I. karategina B. Fedtsch. This appears to be a synonym of *I. lineata*.

I. korolkowii Regel. The most well-known and one of the finest of the Regelia irises. It is 40–60cm in height with the leaves 5–10mm wide and purple-tinged at the base. The rhizome is only slightly stoloniferous and is thick. Each set of spathes produces two or sometimes three flowers and these are about 6–8cm in diameter, rather elongated in appearance because of the long slender sharply deflexed acute falls and strictly erect standards, which are also rather pointed. The colour is creamy white or ivory with delicate blackish-maroon veining and a rather insignificant beard of dark hairs. Variants occur in which the ground colour is purplish, and the veining may be dull green or purple. There is usually a smallish dark signal patch on the falls. Some of these forms have been given names, for example *concolor* which is purple with few veins; *leichtliniana*, cream with few veins and a blackish signal spot; *violacea*, violet veined darker, and *venosa* in which the veins are very prominent. This beautiful species grows wild in Soviet Central Asia in the Tien Shan and Pamir-Alai mountains and in north-east Afghanistan. It flowers in May and June and inhabits open rocky slopes at about 1600–3800 metres.

Planted in a bulb frame, *I. korolkowii* grows well but it is too vigorous a plant for pot cultivation and in my experience does not last long if planted in an unprotected border.

I. kuschkensis Grey-Wilson & Mathew. A very similar plant to *I. darwasica* and *I. lineata*, it differs from them in having very stout rhizomes. The leaves are generally wider than those of *I. lineata*, about 6–8mm wide, and it has broader, rather more rounded falls. It differs from *I. darwasica* in having shorter falls, about 5cm long, and standards about 4–4·5cm long (in *I. darwasica* they are both 6–7cm long).

I. kuschkensis is a robust plant 30–50cm in height with erect leaves, and flowers about 6cm in diameter. These are purple to purple-bronze with deeper veins and a pale purple beard. It grows in the Herat and Paropamisus regions of north-west Afghanistan on grassy hill slopes and sandy gullies at about 1600 metres altitude, flowering in April and often associating with tulips and *Iris fosterana*. It was introduced into cultivation by the Grey-Wilson and Hewer expedition under the number 500 and is probably still in a few specialist collections.

I. leichtlinii Regel. A synonym of *I. stolonifera*.

I. lineata Foster ex Regel. (Syn. *I. karategina*) This is very closely allied to *I. darwasica* and it is possible that the two are variants of one species. It is about 15–35cm in height with short stolons and erect leaves 3–6mm wide. The flowers are brownish-veined or purple-veined on a greenish yellow ground colour and have a bluish beard. The falls and standards are about 4·5–5cm long and are rather narrow and pointed compared to the more rounded broader ones of *I. darwasica*. There are more comments under *I. darwasica*.

I. *lineata* grows wild on dry slopes in Central Asia, in Tadjikistan and adjacent north-eastern Afghanistan. In the latter area it has been collected only once as far as I can ascertain, by Paul Furse (number 8224). A colour photograph of this appears in the *British Iris Society Year Book* for 1968, the form illustrated being rather more purple-blue in overall appearance than those illustrated in earlier works [*Gartenflora* 36, tab. 1244 (1887) and *Botanical Magazine* tab. 7029 (1888)]. In Afghanistan it grows at about 2500 metres on granite slopes near Farkhar.

I. stolonifera Maxim. (Syn. *I. vaga; I. leichtlinii*) A vigorous species, so named because of its habit of producing long stolons from the rhizomes. The erect leaves are about 5–15mm wide and the flowering stems reach 30–60cm in height. There are two or three flowers from each set of spathes and these are about 7–8cm in diameter. The colour is a strange mixture of browns and purple, usually brownish at the margins shading to bright blue towards the centre. The beard is of yellowish or blue hairs. It is a variable species in colour and stature and some of the forms have been given names. These are sometimes available in the nursery trade. In the wild, *I. stolonifera* occurs in Soviet Central Asia in the Pamir-Alai range, flowering in April to June. The *Flora USSR* gives the curious habitat of wet meadows and near mountain streams, but this is not supported by the field notes on specimens I have seen. These suggest a drier more rocky environment at 800–2400 metres altitude, sometimes with spiny cushion plants.

Like *I. korolkowii* it is not a difficult plant to grow but thrives best in bulb frame conditions in Britain.

I. suworowii Regel. A synonym of *I. darwasica*.

I. vaga Foster. A synonym of *I. stolonifera*.

1E   Subgenus Iris, Section Hexapogon

As the name implies, the species of this small group possess a beard on both falls and standards. They are arillate species, that is their seeds have a fleshy attachment, unlike the irises of Section Iris (the 'pogons'), and are therefore more associated with Section Regelia which also contains species with arillate seeds and a beard on all six segments. From the Regelia species the Hexapogons *(I. falcifolia* and *I. longiscapa)* are distinguished by having several flowers included within three or four bracts, whereas in the Regelias there are strictly two flowers (one in *I. afghanica)* enclosed within two bracts. The rhizomes of these two Hexapogon species are very short and knobbly with long, straight, rather fleshy roots, a feature no doubt associated with their habitat, for they are xerophytic plants of the semi-desert regions east of the Caspian, in Russia, eastern Iran, western and southern Afghanistan and Baluchistan. Regelias on the other hand, although geographically nearby, are mountain plants of Afghanistan and adjacent areas of Russia.

Cultivation

These small lilac-flowered desert irises have always been rare in cultivation and are likely to remain so. I know of no plants in British collections at present and on the few occasions I have had the opportunity to try growing *I. falcifolia* it was a case of seeing how long I could prolong its death! From such a specialized habitat it is unlikely that the species will ever take to the changeable English weather conditions very readily. The best advice one can offer is to plant it in a bulb frame with as much sun as possible in summer. Both species should be hardy, for the deserts of Central Asia are bitterly cold, although dryish, in winter. Almost certainly the Hexapogons would do well in some of the hotter, drier states of North America, such as Arizona and New Mexico.

The Hexapogon iris species

I. falcifolia Bunge. A very slender species with a compact non-stoloniferous rhizome and rather fleshy roots. It is usually 10–20cm in height, and forms dense tufts with a cluster of old fibrous leaf bases around the base of the shoots. The leaves are greyish and curved, about 2–4mm wide. From the three or four bracts there are two to five smallish flowers, usually about 3–4cm in diameter. These are lilac-violet with darker veining and a whitish beard on the falls. It is a plant of the semi-desert plains of Central Asia where it flowers in March or April. There are records from a very wide area, from the Kara Kum and Kyzl Kum deserts of Russia, into north-eastern Iran, west and south Afghanistan and south to the Quetta and Baluchistan regions.

It is very closely allied to *I. longiscapa.* This is separated from it by having nearly erect leaves, which are almost rounded in cross-section, and a shorter perianth tube, 2·5cm or less long (about 3cm in *I. falcifolia).* The leaf characters seem to be reasonably reliable but from the herbarium material I have seen, the perianth tube length is not very constant and there is a considerable overlap. The flowers of *I. longiscapa* are possibly slightly larger than those of *I. falcifolia* but even this seems to be marginal. The degree to which the flower is carried above the spathes varies also and it cannot be said that they are always held clear of the spathes, for in some specimens the spathes reach the base of the flower. There has been a misinterpretation by some authors that the long slender perianth tube is a flower stalk, or pedicel. This is clearly not the case, for the ovary is stalkless, held between, and enclosed by, the bracts and it is the perianth tube which pushes the flower up out of its bracts. On the whole, the only difference I can see between the two is the wider curved leaves of *I. falcifolia* which do give the plant a slightly different appearance.

I. longiscapa Ledebour. On the whole the description given for *I. falcifolia* fits *I. longiscapa* except for the leaf characters mentioned above. In *I. longiscapa* they are more or less straight and only 0·5–1·5mm wide. The flower is possibly slightly larger, about 4cm in diameter. The two may well be variants of just one species but unless a thorough study can be made in the wild it is probably best to retain the two as distinct on their leaf characteristics.

I. longiscapa also occurs in the Kara Kum and Kyzl Kum deserts of Soviet Central Asia, but I have seen no records of it from Iran or Afghanistan. It appears therefore to be much more restricted in its range than *I. falcifolia*.

1F   Subgenus Iris, Section Pseudoregelia

An interesting small group of species, and, other than *I. kamaonensis* and *I. tigridia* which are in a few specialist collections, little known as garden plants. Their features are a compact, non-stoloniferous rhizome; flowers with bearded falls which are lilac or purplish and prominently blotched darker; and seeds with a fleshy aril. The combination of characters sets them aside from all the related groups such as Sections Psammiris, Regelia, Oncocyclus and Hexapogon. The standards are unbearded. Mostly they are dwarf mountain plants of eastern Asia but are especially centred on Tibet, Mongolia, Nepal and eastern Siberia.

Cultivation

Apart from the two species mentioned above, little is known of the Pseudoregelia irises in cultivation. *I. kamaonensis* grows and flowers well in my garden in Surrey in a gritty, peaty soil, both in full sun and between heathers. It is however not very long-lived and it is not clear whether this is a natural phenomenon or a failure in cultivation technique. It will also do well in a deep pot, kept only slightly watered during its dormant period in winter, such as under the bench in a cold greenhouse. Its requirements seem to be good drainage, but a cool position with plenty of moisture in spring and summer. *I. tigridia* is less easy and I have twice lost plants after a few years in spite of trying them in a variety of situations from peat banks to bulb frames. The best plant grew in a stony soil in full sun and it flowered and made a clump but at that stage I decided to propagate it and all the divisions failed. An open sunny position, dryish in winter and well-drained but with adequate moisture in spring and early summer would probably suit it best. *I.*

hookerana, *I. kamaonensis* and *I. tigridia* are reported as being the most frequently cultivated species of this group in North America.

The Pseudoregelia iris species

I. goniocarpa Baker. A slender species about 10–30cm in height with erect grassy leaves only 2–3mm wide. The spathes produce solitary flowers and these are flattish and small, about 2·5–3cm in diameter (rarely to 5cm in some specimens from Szechuan), rather variable in colour but mostly lilac or bluish-purple with darker blotching or mottling and a small dense beard of orange-tipped hairs. The falls are rather rounded and the perianth tube is very short. At the base of each of the falls are two gland-like swellings. Variations recorded include an albino, seen in the wild by Reginald Farrer. *I. goniocarpa* flowers between May and July in the wild and grows in a wide area of western China, Nepal, Bhutan, Sikkim and southern Tibet, inhabiting moist grassy meadows and scrub-covered hillsides between 2700 and 5500 metres altitude. I know this plant only as dried material although at one time I had a living plant which failed to flower before disappearing one winter. Dykes mentions growing it from seeds collected by Farrer, the plants doing well for several years in 'light vegetable soil in a sunny garden where it was not allowed to get too dry in the growing season in spring'.

I. gracilis Maxim. The original specimen looks exctly like *I. goniocarpa* (see above).

I. hookerana Foster. A similar plant to the more familiar *I. kamaonensis* but differing from it by having the flowers on short stems (5–12cm) and a perianth tube only 2–3cm long. The flower colour is very variable but usually lilac, purple or bluish with darker mottling, and albinos are also known to occur. It is said to have a sweet scent reminiscent of Lily of the Valley. The species has a more restricted range than *I. kamaonensis* in the western Himalayas, mainly in Kashmir and adjacent parts of India, reaching as far west as Chitral. In the wild it flowers in June or July in subalpine meadows at 2600–4400 metres.

Dykes regards it as a reasonably easy iris to cultivate, recommending a rich light soil which dries out in summer after the foliage has withered. The seeds take a long time to germinate and appear to do so rather sporadically over several years.

I. kamaonensis Wallich ex D. Don. The most well known of the group, with flowers of a rather bizarre appearance. The plant is clump-forming with a thick knobbly rhizome which produces rather fleshy roots and strap-like leaves about 2–10mm wide and up to 45cm long at maturity. The fragrant flowers are nearly stemless but are carried on long perianth tubes about 5–7·5cm long, with the leaves usually slightly overtopping them. They are 4–5cm in diameter and are lilac-purple mottled and blotched darker, so as to have a rather 'virused' appearance! The dense beard is of white hairs tipped with yellow. *I. kamaonensis* is a native of the Himalayas, from Yunnan and Szechuan westwards through Bhutan, Sikkim, Nepal and northern India as far as Simla and into Kashmir. It grows in grass, often amid rhododendrons or bamboos, or in alpine turf at 3000–4700 metres, flowering in May or June. In cultivation it is comparatively easy to grow and in my Surrey garden I have succeeded with it both in a semi-shady bed at the foot of a wall in a gritty-peaty mixture, and in nearly full sun amid heathers in a peat-sand soil. It flowers regularly and is a neat plant at flowering time although the leaves become rather gross later on. Propagation is best by seeds although they rarely seem to be produced in cultivation and cross-pollination between different clones seems to be essential. The capsules are almost stemless at ground level. Sir Colville Barclay and Patrick M. Synge recorded in Nepal a white form which must be a beautiful variant.

I. pandurata Maxim. The original specimen collected by Przewalski in Kansu province, China, appears to be just a vigorous *I. tigridia*.

I. sikkimensis Dykes. This is something of a mystery plant, described by Dykes in 1912 from material thought to have originated in Sikkim. He cultivated it and saw flowers which he noted as being produced on a stem as in *I. hookerana*, but which had a long perianth tube as in *I. kamaonensis*. The leaves were narrower than those of *I. kamaonensis*.

I. tigridia Bunge ex Ledebour. A curious little iris, quite distinct from the other members of the group in its distribution, habitat and general appearance. It makes dense clumps of rhizomes with rather fleshy roots and with fibrous remains of old leaves attached at the apex. The leaves are numerous and erect, only 1–4mm wide and up to 10cm in length, usually slightly shorter than the flowers. The one or two flowers per stem are carried only 10–15cm above ground and are relatively large for the size of the plant, about 4–5cm in diameter. The colour is lilac, mauve or deep blue with darker purple mottling, and there is a yellow-tipped or white beard on the falls. The central area of the blade of the falls, surrounding the beard, is white with purple lines. Dykes mentions yellow-flowered forms but he knew the plant only from herbarium material and may have confused some specimens with *I. potaninii* which has purple and yellow variants. The two are rather similar, but the latter does not appear to have blotched colouring, the old fibres attached to the rhizomes are very markedly curly and the leaf tips are abruptly narrowed to the apex whereas in *I. tigridia* they taper gradually to a point.

I. tigridia is a native of Mongolia, north-western China (Manchuria) and adjacent areas of south-eastern Russia, growing on plains or mountain slopes in grassy steppe, in peaty soils or in gravelly hills at 1400–3000 metres. This curious little species is very rarely cultivated but is in a few specialist collections in Europe and the United States. I have cultivated and flowered it, but cannot claim real success. The best plant was in light stony soil with a good leafmould content and this made a sizeable clump until I tried to propagate by division and lost all the pieces one by one. It obviously dislikes excessive moisture around the crown, especially in winter.

2A Subgenus Limniris, Section Lophiris (The 'evansia irises')

This group of delightful irises are on the whole very distinctive and all of great garden value, although some of them are undoubtedly tender and need greenhouse treatment in Britain. They are a rather diverse group of species, ranging from the tiny North American *I. cristata* to the giant Himalayan *I. wattii*, and the main feature which brings them together is the frilly crest on the falls. This is of a rather different structure from the beard of Oncocyclus and pogon irises and is not really comparable, so that the group has by some authorities been linked with the beardless species rather than the bearded. The cockscomb crest is really an outgrowth from the central ridge or vein of the falls, sometimes with an extra ridge on each side, as in *I. cristata*. In two species the ridge is undissected *(I. tenuis* and *I. speculatrix)* and their inclusion in the 'evansia' group has been suggested, at least in the case of the former, by a combination of other factors. The latter species requires more study before its position is well understood.

In general terms one can say that these North American and eastern Asiatic irises are on the whole creeping plants with widely-spreading stolons and that they tend to be plants of damp woodland or at least mesophytic habitats, not dry-country plants such as are many of the species of Oncocyclus, Regelia etc. Accordingly, they mostly have rather thin or soft leaves, not hard and rigid as in many species, and they are green as opposed to the grey-green appearance caused by a surface layer of waxy 'bloom' which covers the leaves of so many of the more xerophytic ones. Whether or not the species included in this group are in fact closely related to each other one cannot really say and we can only do our best at arriving at some sort of useful classification. Some iris groups fall together very naturally, others less so. It is worth quoting Dr Lee Lenz, from his paper in *Aliso*, **4** (1959), when putting forward the case for transferring *Iris tenuis* into the evansia group: 'From the morphological, ecological and geographical evidence presented it seems best to consider *I. tenuis* as a member of the subsection evansia. . . .' This says a great deal. One must consider as many aspects as possible when coming to a decision about a particular classification, and in the case of the 'evansias' one just cannot say that they differ from all other irises in one or two distinct characters.

The name evansia was produced by Salisbury in 1812 in honour of Thomas Evans of the India House, the introducer of *I. japonica* to cultivation in Britain.

Since this is such a diverse group comments on cultivation methods are given individually for each species.

The evansia iris species

I. confusa Sealy. This is a vigorous clump-forming plant producing short stolons and slender erect stems up to a metre or more in height and looking very like bamboo canes. There are fans of broad leaves at the apex of these canes and in early spring the widely branching flower stems arise from the centre of each leaf fan. There is a long succession of short-lived flowers, each about 4–5cm across, rather flat in appearance and white with yellow and sometimes purple spotting around the yellow crest and signal patch. There are said to be several forms in cultivation differing only slightly in flower colour but I have grown only one. The species grows in the Yunnan province in western China, but its exact distribution is not known and one hopes that Chinese botanists will soon throw some light on this and its relatives.

Subgenus Limniris, Section Lophiris: 1 *I. confusa*; 2 *I. japonica*

Probably most of the material in cultivation derives from that grown by W. R. Dykes; he obtained seeds in 1911 from Père Ducloux, who collected them in Yunnan. Dykes mistakenly called it *I. wattii*, but J. R. Sealy clarified the matter in 1937 by describing the Ducloux plant as a new species, *I. confusa*.

I. confusa is hardy in Surrey where I grow it in a semi-shady place in rather damp soil. It normally flowers in May, but if kept in a slightly heated greenhouse it will begin in late winter. I also grow it as a tub plant for putting on the terrace when in flower. The species just misses being a first-rate garden plant, for its flowers are rather small and the leaves are often brown at the tips and somewhat tattered by flowering time, after the winter weather. The new canes with fresh fans of young leaves are produced in summer, and the old ones pruned out of the clump in much the same way as one prunes raspberry plants! In the United States it is reported as being tender except in the south such as southern California, where it thrives and is obviously a much more attractive plant than when grown in cold-winter climates. It is also very successful in New Zealand and parts of Australia.

I. cristata Solander. A delightful little plant very suitable for rock gardens or peat gardens, forming large patches when growing well. The small much-branching rhizomes produce fans of leaves which are usually less than 15cm long at flowering time but may elongate later, and they are 1–3cm wide. The green spathes produce one or two flowers each and these are about 3–4cm in diameter. The colour varies considerably and some forms have been named. The form normally seen in gardens is a shade of lilac-blue but my American friend Roy Davidson gives the full range as lilac, lavender, blue, purple or violet. In the centre of the falls is a white patch and along this, and down the haft, run three very crisped ridges which have a variable amount of yellow or orange-brown on or around them. The erect standards are much narrower than the falls and are self-coloured. The flowers are almost stemless on the rhizomes and it is the length of the perianth tube itself (about 4–6cm long) which raises the flower well above soil level. *I. cristata* is a native of moist woods in eastern North America, in the Appalachian and Ozark mountains where it grows in more or less neutral soils on the flat or on banks or ledges. Apart from the varying shades of lilac, blue and violet, there are pure white forms which may take the collective name of 'Alba', although there are several different whites on record. One form I flowered had pure white flowers with a yellow central ridge and a faint purplish zone in the centre of the falls. Other forms which have been selected and brought into cultivation include a pink and a very clear blue. I have to thank Roy Davidson of Seattle and Earl Roberts of Indiana for sending me some forms of this beautiful little iris all of which seem to grow vigorously on a peat bank which is sufficiently raised to allow free drainage. The clumps must be lifted and replanted from time to time as they move outwards and away from their original position, dying out in the centre. Division is best carried out soon after flowering.

I. formosana Ohwi. This is something of a mystery plant and one cannot be sure if it is a 'good' species, a hybrid or a form of *I. japonica*. In the description

Subgenus Limniris, Section Lophiris: *I. cristata*

of the species it appears that the larger flowers, and the fact that it sets good seed in the wild, were taken to be two of the main reasons for setting it aside from *I. japonica*. The plant I know under this name was obtained from Maurice Boussard of Verdun who in turn had received it from Mr Chou Cheng in Taiwan. It is a tender, very invasive plant spreading in all directions by means of slender stolons and producing short erect stems up to 10cm in height, some of which terminate in branching inflorescences. The flowers are large and flattish, lilac blue with a yellow crest on the falls. The overall appearance is similar to that of *I. japonica* which, however, has smaller paler flowers and no vertical 'bamboo-like' stems at all. Apart from being a native of Taiwan, very little is known about its behaviour or habitat in the wild. Dr Jack Ellis who has studied this group cytologically has recorded odd chromosome counts in different plants so that there appears to be more than one form around, and there is said to be a little difference in the depth of lavender-blue flower colour between the various forms. The odd chromosome numbers might be accounted for by hybrid origin, but not necessarily so.

As a garden plant *I. formosana* is not very useful in Britain since it is killed by the slightest frosts, even in Surrey. *I. japonica* on the other hand survives unharmed in my garden even through the severest winters. It is reported that *I. formosana* is cultivated by a few specialists in the United States.

I. gracilipes A. Gray. Another of the dwarf species in this section, notable for

its clump-forming, shortly stoloniferous habit and very narrow grassy leaves only about 0·5–1cm wide. The flowers, 3–4cm in diameter, are carried on slender branching stems about 10–15cm tall and are lilac-blue with a large white zone veined with violet in the centre of the falls. The crest is mostly white but is yellow at its apex. One characteristic feature of the species is that it has the bracts fused together at the base instead of being completely separated. This portion sheaths the ovary and base of the perianth tube, which is about 1–1·5cm long. The capsules are also distinctive, being almost spherical. *I. gracilipes* is a Japanese plant, occurring in the mountains of Hokkaido, Honshu and Kyushu and is also present in China. It is a plant of wooded slopes, growing in a leafmould-rich soil and in cultivation does well in the same peat banks where I grow *I. cristata*—i.e. in a situation which is well-drained but not waterlogged and yet is not allowed to dry out in summer. There is a beautiful white form in cultivation which is equally easy to grow. Propagation is by division of clumps during the growing season, or by seed.

I. japonica Thunberg. Probably the most well-known of the evansia irises and one which has been around in cultivation for at least two centuries, it is a vigorous plant, spreading into large patches by means of slender stolons which root down at the apex and produce fans of the broad glossy green leaves which are an attractive feature. There is little tendency to produce erect cane-like stems as in *I. formosana, I. confusa* and *I. wattii* to which it is obviously closely related. The branching flower stems rise to a height of about 45–80cm and produce a long succession of flowers in April and May in Britain. These are flattish, about 4–5cm in diameter and have a delightfully frilly appearance caused by the crisped and fimbriated margins to the falls and standards. The style branches too have wispy appendages at their tips. There is considerable variation in ground colour from white to pale blue-lavender, with an orange crest on the falls, surrounded by a zone of purple blotches. It seems that there are various clones of this plant in cultivation, including one with variegated leaves which is not unattractive as a foliage plant. It is always stated that 'Ledger's Variety' is the best and hardiest in Britain, but exactly how this differs from other *I. japonica* forms is not very clear and it is quite probable that there is no difference between it and 'ordinary' *I. japonica*. There has been a certain amount of hybridization between *I. japonica* and other related species.

It appears that *I. japonica* is a native of Japan and central China, although the exact distribution of a plant which is both horticulturally valuable and which increases rapidly must be in some doubt. There has been much controversy as to the best treatment in gardens and the only advice one can really give is 'try it and see'. In literature one can find recommendations for almost any type of soil and situation. In my own garden it thrives in semi-shade in a rather damp bed at the foot of a peat bank. Most years it has masses of flowers in succession but occasionally it will have an off season. It has never shown severe frost damage, other than brown leaf-tips occasionally.

I. lacustris Nuttall. This really looks like a miniature *I. cristata* and indeed the two are very closely related. Apart from being generally smaller in all its parts, *I. lacustris* has a much shorter perianth tube, usually about 2cm or less. The almost sky-blue flowers (there is also an albino form) have a golden crest and whitish patch on the falls similar to that on *I. cristata*. In leaf width there is perhaps also a generalization to be made; those of the non-flowering shoots of *I. cristata* are 1–3cm wide and those of *I. lacustris* usually less than 1cm. The distribution of the latter is much more restricted in North America and it is confined to the Great Lakes region in moist sands, gravel and limestone crevices, usually in slightly shaded areas at the edges of cedar and fir. In the garden I treat it in the same way as *I. cristata*, planting it on a raised peaty or leafy bed which is well drained but never dries out. The fact that it often occurs on limestone does not seem to mean that it requires a lime-based soil. It flowers in Britain in May and is best propagated soon after flowering, although I have found that pieces carefully detached and potted establish without difficulty at any time.

I. milesii Foster. Unlike the other larger members of this group, *I. milesii* (and *I. tectorum*) produces fat rhizomes which are much more like those of the bearded irises, but are greenish. These give rise to robust sterile fans of pale green leaves 30–60cm in length and about 4–7cm wide. The branched flowering stems reach a height of about 30–75cm and bear shorter leaves than the sterile fans. The flowers have spreading or reflexed falls, often rather folded or undulate at the margins, oblique standards held at an angle of about 45°, and style crests which are fimbriate. The overall colour is a strong pinkish-lavender mottled with deeper purple on the falls, which have a fringed crest of yellow. Each flower is about 6–8cm in diameter and although rather short-lived there is a succession over several weeks. *I. milesii* is recorded over a very wide area of the Himalayas but seems to vary little from place to place. It is an easy garden plant for an open sunny border and sets seeds freely.

I. pseudorossii Chien. An interesting plant which as far as I know has never been in cultivation, it is a small plant of stoloniferous habit, with far-creeping thin wiry rhizomes. It grows only 8–14cm in height, with short stiff narrow grey-green leaves 1–2·5mm wide, barely visible at flowering time. The solitary flowers are about 2–3cm in diameter and can vary considerably in colour. Field notes on specimens state that it can be blue, white or pinkish with a ring of darker colour around a distinct yellow crest. The perianth tube is 3–5cm long, the falls about 2cm long and the standards 1·5–2cm long. *I. rossii* which is of similar general appearance is said to have no crest on the falls and is thus excluded from the 'evansia' group, although some comments about this are given under *I. rossii*.

I. *pseudorossii* was described in 1931 from Kiangsu and Anhwei provinces and has also been collected near Shanghai, Nanking (at the Ming Tomb) and Chinkiang. Flowering time in the wild is in March and it occurs on open hillsides. I have seen the original specimen of *I. proantha* Diels, and can detect no important differences between it and *I. pseudorossii*.

I. pseudorossii var. *valida* Chien. Described at the same time as var. *pseudorossii*, from Chekiang. It is said to have a large flower and to be of more robust habit, up to 28cm in height. The leaves are up to 7mm broad, the falls up to 2·6cm long and the standards 2–2·2cm long.

I. speculatrix Hance. This does not appear to be as mysterious and impossible to cultivate as Dykes indicates in his *Handbook of Garden Irises*. I would however question its position in the Evansia group since it does not really bear much resemblance to any other species and the 'crest' is really just a slightly raised ridge. It seems to me to have more affinities with *I. ruthenica* than with the Lophiris. However, some more work is required on the grouping of irises and there is no point in formally transferring *I. speculatrix* to the Ruthenicae until this is done. It is a tufted or shortly-creeping, grassy-looking plant some 20–25cm tall, with the leaves overtopping the flowers. The leaves, which can be up to a metre or more long, are bright and rather shiny green, 0·5–1cm wide and long-tapering at the apex; at the base they are nearly white with green-veining. Throughout the length of the leaf there is a distinct cross-veining, giving a slightly netted structure. The flower stems carry some reduced leaves and each is two-flowered. The unscented short-lived flowers are about 4–5cm in diameter, and have obovate falls with a blade about 1·5cm wide. They are basically lilac but in the centre of the blade there is a dark violet-purple, almost triangular or heart-shaped zone surrounding a white, purple-speckled area which in turn surrounds the raised yellow ridge. On the haft of the falls there are deep violet veins, and the ridge deepens to orange. The flowers look as if they have a long tube but in fact it is only about 1cm long, the rest of the 'tube' being a solid beak to the ovary. The lilac standards are spreading obliquely rather than erect and are obovate, about 1·5cm wide, narrowed to a short claw at the base. The style crests are long and narrow, about 1·5cm long and 5mm wide. In fruit, the capsules retain their long beak, and are up to 3cm long; they have valves which curl outwards on splitting and contain seeds with a white attachment. The species is a native of Hong Kong, Canton, Lantao Island and Fukien where it grows in grass, scrub or wooded places and flowers in April or May. Understandably it is not hardy in Britain and requires temperate house conditions, but does not seem to be a difficult plant. It is however reported by Jean Stevens (*Iris Society Year Book*, 1950) to grow outside in some New Zealand gardens which are subjected to frosts. Doubtless it would do well in any warm moist climate. The notes given above are mainly taken from plants which I have grown and from those seen at Kew, so that there may be some deviation from these colours and measurementsin the wild populations. The most recent flowering I have seen was of a plant collected by John Simmons, cultivated well by Tony Hall at Kew. It was gathered on Sunset Peak, Lantao Island at 750 metres and flowered at Kew in June 1980.

I. tectorum Maxim. A splendid evansia iris which is instantly recognizable; it has a fat rhizome more like that of a common bearded Iris, and fans of broad thin-textured ribbed leaves, each about 2·5–5cm wide. The flower stems are slightly branched, usually about 25–35cm tall and carry two or three flowers from each set of spathes. The flowers are large, about 8–10cm in diameter, and flattish because the falls are only a little deflexed and the standards are held outwards rather than upright. The ground colour is lilac with darker veins and blotches, especially in the central and lower portions of the falls, and the very dissected crest is white with a few dark spots. The two lobes of the style branches are very raggedly-toothed and these, together with the undulating margins of the falls and standards, give the whole flower a rather frilly appearance. There is a white form, 'Alba', in which the whole flower is pure white except for a few yellow veins around the crest and on the haft of the falls. *I. tectorum* is apparently native in central and south-western China, but is probably not wild in Japan where it is grown on thatched roofs. It is also recorded from Burma. In cultivation it certainly presents no problems if grown in a sunny place, preferably near a warm wall or fence, where it will often spread freely into sizeable patches. It is unfortunately very susceptible to a virus which discolours the leaves badly, but seeds are easily obtained and seedlings are virus-free. Seeds from the albino give white-flowered plants unless they have resulted from cross-pollination with the blue form.

I. tenuis Watson. Largely on the basis of its distribution, this Iris was formerly linked with the Californian group of species. However, Dr Lee Lenz studied the species in great detail and in 1959 suggested that it should be regarded as an evansia. In general appearance it looks a little like a tall *I. cristata*, but produces branched flowering stems 30–35cm in height with small, pale lilac flowers about 3–4cm in diameter. The 'crest' is in fact a low yellowish undissected ridge, not a cockscomb-like structure as in other species. It is a widely-creeping species with long thin leaves about 30cm long and about 1–1·5cm wide. *I. tenuis* has a limited distribution in western North America, occurring only in Clackamas County of Oregon, where it inhabits cool leafy soil in the Douglas fir forests or in dense undergrowth, flowering in May.

I have never succeeded in flowering it and it has been suggested that it is frost-tender in Britain. However, Mike and Polly Stone of Fort Augustus in Scotland report that *I. tenuis* grows and flowers in their garden in peat banks amid primulas and *Meconopsis*, a clear indication of hardiness and the type of growing conditions this iris requires. My own plant grows, not very strongly, in a sheltered peat bed in Surrey and although it lacks the urge to flower it has survived for four years.

I. wattii Baker. There is a considerable amount of doubt as to whether the plant cultivated in Britain under this name is a true species. Dr Jack Ellis has investigated this and other evansias cytologically and is of the opinion that it is a hybrid. A full explanation of the facts can be found in *The Iris Year Book* for 1979 and will not be repeated fully here. It seems that the plants grown as *I. wattii*, in Britain at least, all originate from a collection by Major Lawrence Johnson near the Chinese–Burmese border and represent a sterile, wholly vegetatively propagated, clone. Probably the true story about *I. wattii* and the related *I. confusa* will not emerge until a thorough study can be made of both species (and possibly others as yet unknown) in the wild in Yunnan.

I. wattii, as we know it in gardens, is a magnificent plant but unfortunately tender so that it must be grown in a frost-free greenhouse. Its erect bamboo-like stems reach 1 to 2 metres in height when it is growing really well, with fans of broad leaves and much-branched inflorescences. The flowers are much larger than those of *I. confusa*, about 6cm in diameter, and they are lilac-blue with a whitish area in the centre of the falls, spotted with orange-yellow and darker lilac. The crest is a very prominent ridge and is whitish, also with deep yellow spots. The most obvious feature of the flowers, which makes the species instantly distinguishable from *I. confusa* and *I. japonica*, is that the large falls hang down almost vertically so that they do not have the flattish appearance which the flowers of the other two species have. As far as is known, *I. wattii* occurs in Manipur where it was first collected by George Watt, and in Yunnan province of China at Tengyueh (Tengchung) where the Lawrence Johnson material was gathered. These two areas are separated by about 500 kilometres, and whether or not the species is widespread in this whole general area is not known.

2B(a)   Subgenus Limniris, Section Limniris, Series Chinenses

Unfortunately, probably only one species in this group is in cultivation in Britain, and that is extremely rare. These notes are therefore based mostly on herbarium material, photographs, drawings and descriptions.

The series is not easily defined and the species are in need of much more study in the living state to determine whether they do in fact constitute a distinct group. In certain respects at least some of the species resemble the dwarf evansia irises like *I. cristata* and *I. gracilipes* and it may be that a revaluation of the groups is needed. Without a knowledge of the living plants, however, it is unwise to proceed farther with this, and the Chinenses Series is given here largely unchanged although some of the species listed in the group in P. Werckmeister's *Catalogus Iridis* (1967) almost certainly do not belong and so in this volume they have been moved to other series with which they seem to have more affinity; for example *I. polysticta* will be found in the Spuriae, *I. cathayensis* and *I. kobayashii* in the Tenuifoliae.

The species left in the Series Chinenses are characterized by thin, wiry, usually stoloniferous rhizomes and all occur in China, Korea and Japan. The capsules, where information is available, are markedly triangular in cross-section. The flowers are flattish in appearance because the blade of the falls and the standards tend to spread horizontally. The leaves are prominently ribbed and often appear to be pleated in the photographs I have seen.

The Chinenses irises

I. grijsii Maxim. A shortly-creeping species with apparently tough wiry roots; it grows to about 12–18cm in height when in flower and has prominently nerved leaves up to 20cm long and 4–10mm wide, tapering gradually to an acute apex. The one or two flowers are produced from narrow, pointed spathes and are about 3–4cm in diameter with a tube approximately 1cm long. They are carried on pedicels to 4cm long, measured from the base of the bracts to the base of the slender, grooved ovary. Unfortunately, I have not traced any fruiting material to determine the shape of the capsule, and colour notes for the flowers are not available. *I. grijsii* is recorded from south-eastern China in the Fokien region where it flowers in May, apparently at altitudes around the 1800–2000 metres level.

I. henryi Baker. As with most of the series, this interesting-looking species is not known to be in cultivation. It is a very slender plant about 12–15cm in height with thin, wiry, creeping rhizomes looking almost like those of couchgrass. The leaves are often only 1–2mm wide and overtop the flowers at flowering time. In their narrowest form they look almost crocus-like and, from dried material, there is the suggestion of a pale stripe along the centre. However, some specimens show that they can be 4–5mm wide and prominently ribbed. The 3cm long narrow, green, spathes produce two flowers on slender pedicels 3–5cm long and these have a very short tube only 2–3mm in length. They are described as being white or blue and appear (from herbarium specimens) to have a diameter of about 3cm, rather rounded blades to both the falls and the shorter standards, and very small style branches. *I. henryi* is a native of central China in Hupeh province and is said to occur in mountains at about 1800 metres.

In several of its characters it resembles *I. odaesanensis*, but the much narrower leaves enable the two to be clearly distinguished.

I. koreana Nakai. This appears to be a very similar plant to *I. minutoaurea* and it seems very likely that the two are just forms of one species. *I. koreana* is described as being a taller more robust plant of loose clump-forming habit with slender rhizomes and long stolons. The leaves are up to 1·3cm in width (only 2–3mm in *I. minutoaurea*) and 35cm long, with 10–14 prominent veins. The flower stem is shorter than the leaves and usually produces two flattish flowers, caused by horizontally spreading falls, probably about 3–4cm in diameter. They are predominantly yellow with a brownish stain on the haft of the standards. The obovate falls bear slightly raised and crinkled ridges on their hafts while the suberect, much paler standards have an elliptical blade notched at the apex, and a very narrow haft. The style branches are the same pale creamy colour as the standards. Like *I. minutoaurea*, it has a long perianth tube exserted from the narrow green bracts.

I. koreana is, as its name informs us, from Korea, in the central and southern regions, inhabiting dry or dryish places in scrub. It is not apparently in cultivation in Britain.

I. minuta Franch. & Sav. A synonym of *I. minutoaurea*.

I. minutoaurea Makino. (Syn. *I. minuta*). A tiny plant of obscure origin although almost certainly a native of Korea or China. It is much-cultivated in Japan and Korea and it may in fact be a dwarf form of *I. koreana*, although this cannot be proved and for the purposes of horticulture the two are distinct. The plants reach only 8–10cm in flower and are clump-forming with thin wiry rhizomes, and they give rise to tufts of leaves only 2–3mm wide. The flowers are only 2–2·5cm in diameter and have tiny spoon-shaped falls and even smaller standards. As with *I. koreana*, they are yellow with paler standards and style branches, and the hafts of the standards have a brown stain. The spreading falls and style branches give the flower a flattish appearance, although it does have a perianth tube about 2·5cm long. In cultivation in Britain it flowers in March to May, if one is very lucky! There seem to be no special cultural recommendations that one can give, although the species does not seem to thrive on my peat bank for long, so that situation is probably best avoided. The general opinion from literature seems to be that it likes a heavy rich soil and plenty of moisture in spring, more or less in full sun. I have seen it flower only once in about 15 years and have in any case now lost my plant.

A plant which was 'going the rounds' as *I. minutoaurea* for some years in the 1960s and '70s eventually flowered and turned out to be a species of *Acorus*. The foliage of the two is similar and neither is free-flowering in Britain, so the mistake was nearly forgiveable!

Although the name *I. minuta* was used before *I. minutoaurea*, and therefore should have priority, there is an even older use of *I. minuta* by Linnaeus, who was actually referring to a species of what we now call *Moraea*. The re-use of *I. minuta* by Franchet and Savatier in 1879 is not allowable under the International Rules of Nomenclature and Makino accordingly provided the new name *I. minutoaurea*.

I. odaesanensis Lee. A recently described species, in 1974, from Korea, it is allied to *I. koreana* but has white flowers, two per stem. The rhizomes have long slender stolons and it is a plant of about 20cm in height. The leaves are long-tapering at the apex, just over 1cm wide and rather short at flowering time but elongating later to 12–35cm at maturity. The two white flowers appear to be flattish because the falls and standards are held out horizontally. They are carried on long pedicels, are about 3–5cm in diameter and have a very short tube. The falls are obovate and rather blunt or rounded and there is a yellow signal patch in the centre, while the standards are slightly narrower and more lanceolate. The drawing and photographs available show that the three style branches are rather small with acute lobes and it also looks as if there is a frilly ridge on the haft of the falls. The capsules are 2–3cm long, very markedly triangular (in cross-section), about 1–1·3cm wide and are dangling on long pedicels which are well exserted from the bracts.

This species takes its name from Mount Odaesan in Korea where it flowers in May. Unfortunately I have no notes about its habitat but the colour photographs and general habit suggest a woodland environment. It appears from a drawing to be like *I. henryi* in its long thin wiry rhizomes and in the flowers having only a very short (less than 5mm) perianth tube. However the much broader leaves, shorter than the stem at flowering time, give *I. odaesanensis* a rather different appearance. It is illustrated in the beautiful *Illustrated Flora and Fauna of Korea*, Volume 18 (1976) by Dr Yong No Lee.

It would be interesting to study this iris in its living state, not only for its obvious beauty but to enable an opinion to be formed as to its relationships. The suggestion of a ridge or ridges on the falls may indicate an affinity with the

Lophiris (evansia) group rather than Chinenses.

I. rossii Baker. This is named after John Ross who collected plants in northern China in 1876. It is a slender species 10–15cm in height, rather tufted in habit, with tough rhizomes which are only shortly stoloniferous, if at all, and somewhat bristly around the base of the shoots. The thin grassy leaves are about 2–5mm wide and are glaucous on one side. They are shorter than or equalling the flower stem, elongating later to about 30cm. The stems are single-flowered but clumps often have several stems and therefore appear quite floriferous. Each flower is 3–4cm in diameter and is flattish because the falls and the standards both spread out more or less horizontally. The falls are obovate or elliptical, abruptly narrowed to a short haft, while the slightly shorter standards are also obovate but narrowed more gradually to a more slender haft. Although *I. rossii* is stated to have no crest, in one of the colour photographs I have seen there is the suggestion of a raised crinkled ridge on the haft. The flowers have a perianth tube 4–7cm long and are produced from long (6–7cm), gradually tapering bracts which are green, and at a glance they are very similar to the equally acute-tapering leaves. The colour of the flowers varies but is usually in purple or violet shades with a whitish haft to the falls, which is spotted and veined violet with a slightly yellowish stain in the centre. Ross described them as being lilac to pink, and he saw one white form. The notes on a Carles specimen from Seoul indicate yellow, but there may be some confusion here. A gorgeous white form with a yellow tinge on the haft of the falls has been described as forma *alba* by Yong No Lee of Seoul. This also appears to have a ridge on the falls and one wonders if *I. rossii* truly possesses a crest or whether some of these (I have seen photographs only) should be referred to *I. pseudorossii*. Apart from the alleged lack of a crest, *I. rossii* differs from *I. pseudorossii* in being a larger, more leafy plant of tufted habit, whereas the latter has thin widely-spreading stolons and is of a much shorter stature. These features are not easy to ascertain in the available photographs.

    I. rossii grows on dry grassy banks and in scrub in northern China, Korea and in Japan on Honshu, Shikoku and Kyushu. It flowers in April or May in the wild and produces rounded stemless capsules in July. As far as is known the species is not cultivated in Britain.

2B(b)   Subgenus Limniris, Section Limniris, Series Vernae

A dwarf North American species, *Iris verna* is somewhat out on a limb and the only representative of its series. Like the rest of Subgenus Limniris, it is a rhizomatous species with no beard on the blade of the falls but is distinguished from all the other series by the three-cornered capsules which contain rounded seeds with a fleshy appendage (this soon dries out after leaving the capsule). *I. ruthenica*, another beardless iris which has seeds with a white appendage, is quite a different plant. Its capsules are small and rounded rather than three-cornered and it is an Asiatic steppe plant whereas *I. verna* is essentially a woodland species from the south-east United States.

I. verna Linn. This is a miniature iris, only 4–6cm tall when in flower, but the leaves can eventually reach 10–15cm in length. The rhizome is rather compact and clump-forming, or sometimes fairly widely spreading, and produces small fans of greyish-green or deep green leaves each about 3–13mm wide. The flowers are 3–5cm in diameter, carried on very short stems but they have a perianth tube 2–5cm long. The colour is a bright, rather clear lilac-blue with a narrow orange stripe in the centre of the falls. It is a variable plant, sometimes recognized in two varieties; var. *verna* in which the rhizomes are slender and 5–15cm long between the sideshoots—this has leaves 3–8mm wide and a capsule 1·2–1·8cm long—and var. *smalliana* Fernald, with short thicker rhizomes only 1–3cm between sideshoots and leaves 6–13mm wide; the capsules are 1·7–2·3cm long.

I. verna flowers in March to May in the wild and inhabits rocky woodlands on acid soils in the south-eastern States of North America, in the Appalachian and Ozark mountains, extending eastwards to coastal dune country.

In Britain it is rather a rare plant, although apparently not very difficult to grow. Thanks to Roy Davidson of Seattle I have had the opportunity to try it and I find that it grows well on a raised part of the peat garden in a position where it does not dry out, but at the same time is never waterlogged.

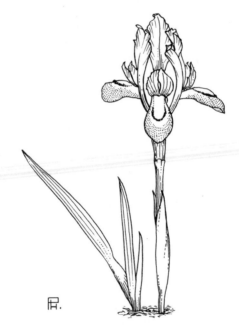

Subgenus Limniris, Series Vernae: *I. verna*

2B(c)   Subgenus Limniris, Section Limniris, Series Ruthenicae

Although there is probably only one species in this series, it is exceedingly variable and quite a number of names exist for the more extreme forms. From a horticultural point of view some of these variants are really quite distinct and it would be a pity to lose track of them by lumping them under *I. ruthenica*. It is probably best to recognize just one species for taxonomic purposes but at the same time say that there are names such as *I. uniflora* available, if one wishes to distinguish the particularly dwarf garden-worthy variant which is in cultivation.

The feature which distinguishes the Ruthenicae from other members of Section Limniris is the small (1–1·5cm long) rounded un-ribbed capsule containing teardrop shaped seeds which have a fleshy appendage. The three valves of the capsule open quickly and curl back releasing their seeds over a short period. *I. ruthenica* is entirely Asiatic, unlike the structurally rather similar (but not in overall appearance) *I. verna* (Series Vernae), which is North American.

I. caespitosa Pallas ex Link. The plant I have grown under this name came from Dr Rokujo of Japan. It produces clumps up to 20cm across consisting of a mass of narrow grassy leaves 15cm or less long, amid which are carried small solitary deep violet flowers in May. It is an excellent plant for a sunny part of the peat garden, but the flowers have a curious smell somewhat reminiscent of a disinfectant or of *Hamamelis japonica*.

I. ruthenica Ker-Gawl. A tufted iris with a shortly-creeping rhizome and erect grassy bright green leaves up to 30cm in length and about 2–5mm wide. The fragrant flowers are carried on short stems 3–15cm tall and are usually solitary, or sometimes two in each set of bracts, and are 3–4cm in diameter with a perianth tube about 1cm long. The falls are white with blue-lavender or violet margins and veining and the standards and styles are wholly violet or bluish-lavender. Usually the blade of the falls is speckled in the centre and it is held out horizontally. There is no prominent ridge on the falls, or any trace of a yellow signal stripe. The standards and style branches are held well up, nearly to the vertical position, and are rather prominent. The capsules remain on short stems amid the leaves. It is a very widespread plant in eastern Europe and Asia from Roumania, east through Central Asiatic Russia to China and Korea. Its habitat is on dry hills and grassy plains, at the edge of woods and in dryish pine and birchwoods, from low altitudes up to 3200 metres. It is therefore not surprising that this is a variable species! The flowering period is normally May to July.

Some forms have nearly stemless flowers among the leaves which are only 5cm long and 1mm wide (e.g. var. *nana*).

All the forms I have grown seem to be easy garden plants requiring only a soil which does not dry out excessively in summer, and is not waterlogged in winter. It and its variants are good plants for the rock garden.

I. ruthenica var. **brevituba** Maxim. As its name suggests, this is a variant with a short perianth tube.

I. speculatrix Hance. This is described under the Lophiris group (page 75) but, as mentioned there, it may have affinities with *I. ruthenica*.

I. uniflora Pallas. The *Flora of the USSR* separates this as a distinct species, distinguished from *I. ruthenica* by having broader leaves (0·8–1cm) and the stem leaves close to the flowers rather than on the lower part of the stem as in *I. ruthenica*. Its distribution is from eastern Siberia to the far eastern provinces of Russia. There is a beautiful albino, var. *alba*, in which the whole flower is white with no trace of colour, even in the centre of the flowers. This came to me from Roy Davidson as did several other of my 'special' irises. Both this white form and *I. uniflora*, as I grow it, are very dwarf and similar to the *I. caespitosa* described above.

2B(d)   Subgenus Limniris, Section Limniris, Series Tripetalae

This Series is so called because the standards are reduced, often to bristle-like proportions, so that the flowers appear to be 'three-petalled'. Within the Subgenus the two species described below are the only ones to have this feature although it does occur in other parts of the genus, as in *I. danfordiae, I. serotina* and several juno irises.

    I. setosa has one of the widest distributions in the genus, from eastern Asia through Japan across to Alaska then jumping to the eastern side of Canada. *I. tridentata*, on the other hand, is restricted to the south-eastern United States.

Cultivation

I. setosa is an extremely easy iris to grow. The tall forms will grow in any good garden soil if well supplied with water and lime-free. The smaller variants are very attractive as rock-garden plants. *I. tridentata* is reputed to be tender although I have not tried to grow it. It grows in rather damp places and will probably be most satisfactory in mild, moist climates.

The Tripetalae irises

I. setosa Pall. ex Link. A thoroughly good garden plant, and very variable so that one may have several horticulturally quite distinct plants belonging to the same species. Some of the variants have names and appear to breed true from seed if self-pollinated. It varies in height, about 15–90cm, from fairly stout rhizomes which are often clothed with the fibrous remains of old leaf bases. The leaves are often reddish-tinged at the base and vary from 20–50cm in length and 1–2·5cm in width. The stem is usually much-branched, although in the very dwarf forms may be simple, and it carries green bracts which are often purple-margined. There are solitary-flowered variants but normally the bracts enclose two or three flowers which are 6–9cm in diameter. The perianth tube is about 5–10mm long and about the same width as the ovary so that the two look almost continuous at the peak of flowering time, although if fertilized the ovary soon swells and gives rise to bluntly three-angled capsules. The colour varies considerably, but in the Asiatic and north-western American forms it is usually purple-blue while in those from the eastern states of Canada it is often a clearer blue. The narrow haft of the falls is veined purple and blue on a pale yellowish or whitish ground and it expands abruptly into the large

SUBGENUS LIMNIRIS, SERIES TRIPETALAE
1. *I. setosa*
2 *I. setosa* subsp. *canadensis*

orbicular or ovate-elliptic blade. The standards are at the most 2cm in length and are very narrowly lanceolate with a sharply acute point, or they sometimes have two lobes at the base. For the size of the flower, the style branches are comparatively short and insignificant, each with two small but broad, toothed, lobes. *I. setosa* occurs in wet meadows, in peaty bogs, in light woods and on shores, usually at rather low altitudes and it flowers in June to August. It is recorded from eastern Siberia and the eastern seaboard of Russia from Kamchatka south to China and northern Korea, in the Kurile Islands, Sakhalin and the Aleutians and in Japan, on Honshu and Hokkaido islands. On the North American continent it is in the west in Alaska and in the east in Labrador, Newfoundland, Prince Edward Island, New Brunswick, Quebec, Ontario and Maine.

The following variants have been named, and some of the plants are certainly very distinct and are stable when propagated from seed. The nomenclature is obviously in some chaos and is in need of attention. The epithets are here deliberately arranged in alphabetical order so that there are no taxonomic implications!

I. setosa forma **alpina** Komarov. A very short-stemmed form from Siberia.

I. setosa var. **arctica** (Syn. *I. arctica* Eastwood). A dwarf variety from Alaska with purple flowers, variegated white.

I. setosa subsp. **canadensis** (M. Foster) Hulten (Syn. *I. hookeri*). This is a dwarf form from the eastern States of North America with an unbranched, nearly leafless stem and usually solitary (rarely two) flowers. The colour is normally a more lavender blue than that of the Asiatic and Alaskan variants. Even this is variable, however, and in turn some variants of it have been given names.

I. setosa subsp. **hondoensis** Honda. A robust plant from Japan with large purple flowers on stems up to 75cm tall.

I. setosa subsp. **interior** (Anders.) Hulten. This is an Alaskan plant having shorter, more papery, violet-coloured bracts and narrow leaves.

I. setosa var. **nasuensis** Hara. A robust Japanese form up to 100cm in height with wide leaves and very large flowers which are more like those of an *I. laevigata*, although the standards are reduced considerably in size.

I. setosa forma **platyrhyncha** Hulten. A form from Alaska with solitary flowers and larger more expanded standards than is usual for the species. I have seen one plant of *I. setosa* in which the standards are nearly as large as the falls but this seems to be an infrequent abnormality.

I. setosa forma **serotina** Komarov. Another Siberian form, described as having solitary stemless flowers.

In addition to these there is a splendid white form, given to me by Roy Davidson. This grows to about 60cm in height, has made large clumps and is very free-flowering.

I. tridentata Pursh (Syn. *I. tripetala*). This grows to about 30–70cm in height from fairly slender tough rhizomes and has leaves 3–10mm wide, shorter than the flowers. On the stems, which are simple or with one lateral branch, there are reduced leaves. The one or two delicately scented flowers are about 5–10cm in diameter; the falls have a nearly orbicular blade 2·5–3·5cm in diameter, which is violet or bluish-purple veined darker, with a whitish signal area and yellowish stain, or velvety yellow blotch, in the centre. The haft is white with brownish netted veining. The perianth tube is about 2–2·5cm in length and the violet standards are erect and oblanceolate, only 1–1·5cm long. The style branches are quite small in relation to the rest of the flower. In fruit, *I. tridentata* has an obscurely angled capsule with flattened semi-circular seeds. It is a plant of damp grassy areas and ditches from the south-east of North Carolina to Florida, flowering in May and June. I have not had the opportunity to grow this and therefore cannot comment on its cultivation.

I. tripetala Walt. A synonym of *I. tridentata*.

2B(e)   Subgenus Limniris, Section Limniris, Series Sibiricae

This well known group of garden plants, containing the Siberian iris and its relations, is sometimes split into two, the Sibiricae consisting only of *I. sibirica* and *I. sanguinea* and the rest placed in a group with *I. chrysographes*. This division may well be correct, based on their cytology, but morphologically it is not easy to justify further splitting of the beardless irises.

The Sibiricae are all Asiatic plants, mostly from central and eastern Asia, but *I. sibirica* extends westwards into central Europe. They are on the whole fairly tall plants liking damp conditions in mountain meadows or light woodland and have stout rhizomes producing hollow-stemmed (except *I. clarkei*), simple or few-branched inflorescences. The capsules are three-cornered, sometimes nearly rounded in section, and they contain D-shaped or nearly cubical seeds. In common with the Californian group, the species of the Sibiricae have a stigma flap which is like a triangular tongue, although there is no reason to confuse the two series in any other way.

They are all deciduous, the above-ground parts dying away completely for the winter months.

Cultivation

All the species in the series are easy garden plants in Britain, requiring only reasonably rich soil which does not dry out in the spring and summer growing period. They are ideal border plants and respond well to the dressings of well-rotted compost which one gives to the other herbaceous border plants in autumn.

There is a useful account of the Sibiricae by Dr C. Grey-Wilson in a booklet published by the British Iris Society in 1971.

The Sibiricae irises

I. bulleyana Dykes. This is a plant of somewhat dubious origin and it seems highly probable that it is a hybrid of *I. forrestii* and *I. chrysographes*. Dykes obtained the material for his description from A. K. Bulley who said that he had raised it and *I. forrestii* from seeds collected by George Forrest. However, Forrest denied seeing any plant like it in the wild and the evidence suggests that it is a natural hybrid, for there was not sufficient time between the first introduction of *I. forrestii* and the appearance of *I. bulleyana* for *I. forrestii* to have hybridized and produced flowering offspring in gardens. If *I. bulleyana* is self-pollinated it apparently produces a variety of forms, some rather like *I. forrestii*. Since in the wild this and *I. chrysographes* occur in the same regions it is possible that natural hybridization is taking place.

In the form of *I. bulleyana* illustrated by Dykes, the hollow stems are unbranched, about 35–45cm in height. The flowers are about 6–8cm in diameter with rather spreading falls which are bright violet at the tips, becoming variegated with deep violet dots and streaks on a white ground in the centre and shading to a greenish-yellow on the haft. The standards and style branches are a more uniform violet.

I. chrysographes Dykes. A beautiful garden plant, perhaps the best in the series for its delicate flower markings, it grows about 35–45cm in height with unbranched hollow stems and grey-green leaves about 1–1·5cm wide. The two fragrant flowers are about 6–7cm in diameter and have falls which droop downwards to the vertical position. The overall colour is a beautiful reddish violet, broken by gold streaks in the centre of the falls but there are variations in which the gold lines are absent, and some forms are a deeper reddish colour than others. Var. *rubella* is a particularly fine deep form which is in common cultivation. *I. chrysographes* occurs wild in China, in Yunnan and Szechuan provinces and in north-east Burma at varying altitudes (1300–4500 metres) where it flowers between July and September. Its habitat is in moist pastures and marshes.

In cultivation it is a very easy and superb garden plant which hybridizes with *I. sibirica* and *I. forrestii*.

I. clarkei Baker. A widespread species in the wild, particularly in the eastern Himalaya, but is unfortunately not seen a great deal in British gardens. It differs from the rest of its Series in having solid stems, but otherwise looks as if it should belong here. The leaves are about 1·3–2cm wide, rather broader than in the other species. The stems reach about 60cm in height and frequently carry one to three branches, each of which is usually two-flowered. The flowers are about 7–7·5cm in diameter and look a little different in shape to other Sibiricae species because the standards are curved out to the horizontal position. The colour varies from a mid blue-violet to dark blue or reddish-purple, with a large white, violet-veined signal patch in the centre of the falls. On the haft there is a little yellowish coloration. It flowers in the wild from May to July in central and eastern Nepal, Sikkim, Bhutan, Tibet and the Manipur region of north-east India and grows at altitudes of 2500–4300 metres on hillsides in damp grassland, sometimes at the edge of rhododendrons and abies forest. In some places it is said to be so plentiful that the dried leaves are gathered like hay and fed to horses and yaks, although John Templar, who knows the plant well in the wild, informs me that they do not graze on it whilst it is in the growing state.

I. delavayi Micheli. A robust Sibiricae, reaching to 1·5 metres in height when in flower, with the grey-green leaves considerably shorter, usually up to 90cm. The stems may have one to three branches, each branch bearing two large flowers about 7–9cm in diameter. These are variable in colour from light to dark purple-blue with a sizeable white signal area in the centre of the wide, rounded blades of the falls. The standards are inclined at an oblique angle rather than erect as in *I. sibirica* and *I. sanguinea*. It is a native of boggy pastures in the Chinese provinces of Western Yunnan and south-western Szechuan, flowering in July or August at altitudes of 3000–4000 metres. In gardens it grows well if given almost bog-like conditions, or at least a lot of water in the growing season.

I. dykesii Stapf. This is rather a mystery plant since it is of unknown origin, first flowering in Britain after Dykes had died. C. Grey-Wilson postulates that it might be of hybrid origin, as in *I. bulleyana*. The plant is like a vigorous *I. chrysographes*, with leaves up to 2cm wide, grey-green and with the unusual feature of sheathing the stem for most of its length. The flowers are a bright deep violet-purple with white and yellow veins in the centre of the falls, and are somewhat larger than those of *I. chrysographes*. *I. dykesii*, as figured in the *Botanical Magazine*, tab. 9282 (1932) is possibly not now in cultivation although plants under this name are seen from time to time.

I. extremorientalis Koidzumi. A synonym of *I. sanguinea*.

I. forrestii Dykes. A slender species about 35–40cm in height, but sometimes as dwarf as 15cm in the wild. The narrowly linear leaves are glossy on one side and grey-green on the other and are much shorter than the flower stems, which are unbranched with a terminal inflorescence of two flowers. The scented flowers are only 5–6cm in diameter, and are clear yellow with brownish-purple lines on the haft of the rather narrow falls. The standards are erect whereas in the larger-flowered yellow *I. wilsonii* they are held at an oblique angle. *I. forrestii* was collected and introduced by George Forrest in 1908 from western Yunnan in China but it also grows in south Szechuan and northern Burma. It is a plant of alpine pastures at about 2900–4300 metres and flowers in June in the wild.

In gardens it is not difficult to grow, but will hybridize with other purple-flowered species to produce some strange intermediate colours.

I. orientalis Thunberg. A synonym of *I. sanguinea*.

I. phragmitetorum Handel-Mazzetti. This species is known to me only as a dried specimen and from the original botanical description. It is a plant at least 50cm in height when in flower, with shorter leaves about 45cm long and 1–1·5cm wide. It has unbranched stems producing large terminal flowers,

SUBGENUS LIMNIRIS, SERIES SIBIRICAE: 1 *I. forrestii*; 2 *I. sibirica*

probably about 8–9cm in diameter, dark blue with a white, blue-veined area in the centre of the falls. It was described in 1925 from material collected in reed swamps in the north-western part of Yunnan province of China, where it was in flower in April. Living material is needed in order to establish its relationships with the Sibiricae Series.

I. sanguinea Hornem. ex Donn. (Syn. *I. orientalis* Thunb., *I. extremorientalis*, *I. nertschinskia*). A far-eastern species much confused with the more westerly occurring *I. sibirica*. They are very similar but have certain distinguishing features which are fairly easily seen in the true plants—unfortunately there are garden hybrids between them which obscure the differences. *I. sanguinea* is variable in stature in the wild, 30–75cm in height, and has slightly glaucous leaves 5–12mm wide, usually about equalling the unbranched flower stems. There are two or three flowers in each set of green bracts and they are reddish-purple, about 6–8cm in diameter, carried on more or less equal pedicels. The falls have a broadly ovate or nearly orbicular blade with a yellowish or nearly orange haft, finely net-veined with purple. A white form occasionally occurs in the wild. The standards are quite small and erect, and the stamens are reported as having dark purple anthers although I have not checked this feature on true wild collected material. In fruit, it has elongated capsules, three times as long as broad, carried on roughly equal pedicels. *I. sanguinea* is a plant of mountain meadows, open wet plains and light deciduous woods, flowering in May, June or July. It occurs in south-eastern Russia—as far as I can ascertain, from the east of Lake Baikal eastwards to the Ussuri region, into Korea and Japan.

'Alba' is the white form, usually with some purple veining.

'Violacea' has deep violet flowers with larger standards than the usual form. It is said to have originated in Korea.

I. sanguinea is distinguished from *I. sibirica* in having the leaves equal to or even longer than the unbranched flower stems (leaves shorter than the branched stems in *I. sibirica*), reddish-purple flowers (blue in *I. sibirica*) which are rather large and produced from bracts which are green at flowering time (brown in *I. sibirica*). The two or three pedicels are usually roughly equal in length in *I. sanguinea* whereas in *I. sibirica* they are markedly unequal; this shows up especially at fruiting time when they have elongated somewhat. At this stage a further difference becomes clear; the capsules of *I. sibirica* are about twice as long as broad, thus appearing more squat. *I. sanguinea* crosses so readily with *I. sibirica* that one has difficulty in maintaining the true stocks other than by vegetative propagation.

I. sibirica Linn. Much has been written about the relationship of the Siberian iris with *I. sanguinea*, and need not be repeated here. This well known and fine garden plant has branched stems 50–120cm in height, well-overtopping the narrow leaves which are usually green and up to 4mm wide. The blue or blue-violet dark-veined flowers are about 6–7cm in diameter, two or three and sometimes more being produced from each set of brown papery spathes. The falls have an obovate or oblong blade with a white, strongly violet-veined area in the centre, narrowing to a paler haft which has prominent dark veining. The standards are smaller and more uniformly blue-violet. In fruit, the capsules are rather tubby, ellipsoid and not more than twice as long as broad, carried on very unequal stalks. *I. sibirica* has a more western distribution, from northern Italy eastwards through central and eastern Europe to Russia, extending to the west of Lake Baikal. East of Lake Baikal, *I. sanguinea* takes over and occurs eastwards to Japan. *I. sibirica* also occurs in the Caucasus, and I have collected it in north-eastern Turkey. Like *I. sanguinea* it is a plant of wet meadows, damp forest margins and light deciduous woodland, flowering in May to July.

There are of course many cultivars available from white through all shades of blue to deep violet-blue, all extremely good garden plants for the herbaceous border or pond or stream margins. If grown away from water they thrive perfectly well if supplied with plenty of moisture during the growing season.

I. wilsonii C. H. Wright. This grows to about 60–75cm in height and has grey-green leaves (on both sides) about equalling the flowers stems which are unbranched and carry two fragrant flowers. They are about 6–8cm in diameter, pale yellow with the broad obovate falls veined and dotted with purple-brown, especially in the centre and on the haft. The standards are held at an oblique angle, not erect as in the smaller-flowered *I. forrestii*, and in the fruiting stage it can also be separated from this species by having elongated pedicels up to 10cm long (not more than 7·5cm in *I. forrestii*). *I. wilsonii* was collected by E. H. Wilson in 1907 in China and is known to occur in the provinces of western Hupeh and western Yunnan, and probably also in Shensi and Szechuan. It occurs at altitudes of 2300–4000 metres on rocky hillsides on the margins of streams and in alpine meadows, flowering in July.

Being yellow-flowered it is distinct from all other Sibiricae except *I. forrestii* and can be separated from this by the large flowers with oblique standards, leaves which are glaucous on both sides and long fruiting pedicels. Unfortunately, as with *I. forrestii* it crosses with other species and loses its identity unless great care is taken.

2B(f)   Subgenus Limniris, Section Limniris, Series Californicae ('Pacific Coast Irises')

A beautiful group of irises, the 11 species in all which are popularly known as the 'Californian group' or 'Pacific Coast Irises', do in fact extend outside California into Oregon and Washington. The group has been thoroughly revised by Dr Lee W. Lenz of the Rancho Santa Ana Botanic Garden and this is published in the journal *Aliso*, **4** no. 1 (1958). Additional information is available in the British Iris Society's publication, *A Guide to the Pacific Coast Irises* by Victor A. Cohen (1967) where the author's own observations are added to the taxonomic treatment by Dr Lenz.

There is obviously little to be added to these excellent pieces of work and I have therefore made no attempt to study the Californian group taxonomically.

I have seen only a few of the species in their wild state and only one of these was in flower, so many observations are from herbarium specimens. I have however grown several of them from wild collected material and as living plants in the garden I have a knowledge of about half a dozen species. The following notes are based on the above source.

The Californian irises are distinguished as a group from other species in the various beardless series on a variety of characters including their distribution, in the Pacific Coast States, and their chromosome count (2n=40). In their morphological features they are on the whole recognizable by their tough rhizomes with wiry roots, producing usually slender leaves, and mostly unbranched stems. On the whole they are fairly dwarf plants producing one or two flowers per stem, although some species have three or four. The flowers are of typical iris form, although on the whole the blades of the falls tend to spread out horizontally rather than being down-turned sharply. The stigma flap on the underside of the style branches is often a triangular tongue, but is squared-off at the apex or somewhat bilobed in *I. purdyi*. In cultivation in southern Britain most of them seem to be evergreen but Dr Lenz says that this is not the case in the wild, where some of the higher altitude species are definitely deciduous. The capsules are often triangular in cross-section, but some are nearly rounded. The Californian irises are ecologically mountain plants, occurring in neutral or slightly acid, gritty, well drained soils in lightly wooded areas, although *I. douglasiana* is often found in the open on coastal headlands and in heavier soils. The species are mostly extremely variable, especially in flower colour. In addition to this, identification is complicated by the natural hybridization of most of them. In gardens they will cross freely, sometimes producing beautiful offspring and they have also been hybridized with the Sibirica group to produce the 'Cal-Sibe' irises. Many of these have been given cultivar names, but to increase any particular form or hybrid it is advisable to propagate vegetatively since they are likely to produce a mixture from seed.

Cultivation

In Britain, most of the species are very easy to grow in well-drained, slightly acid soil in full sun or dappled shade. They are ideal plants for raised peat beds which are not too shady. *I. douglasiana* is by far the easiest and most vigorous and will thrive in many different situations, sun or shade, heavy clay or light sandy soils, acid or slightly alkaline. Some of the species are undoubtedly tender, even in Surrey, and I have lost *I. macrosiphon* and *I. tenuissima* in relatively mild winters. *I. chrysophylla* is also reported as being tender and *I. munzii* is the least frost-resistant, really a subject for the frost-free greenhouse. Adequate moisture is required in spring and this is the period, just after flowering, which is probably the best time to lift and divide the clumps since new roots are being produced. In summer the plants will tolerate dry conditions if well established, but it is not the time to move or propagate them. Seeds are freely produced and germinate readily, but if more than one species of the group is growing in the garden then hybridization is quite likely to have taken place. Seedlings are best potted up and kept in a frame for the first winter, then planted into their permanent position in the spring. The Pacific Coast irises mostly flower in April, May or June in Britain.

In distinguishing between the species it is important to know the length and shape of the perianth tube and the appearance of the bracts, which may be held close together around the pedicel and perianth tube or splayed well apart. The flower stem may be clothed with reduced bract-like leaves or bear fewer, more typical leaves. The flower colour varies a great deal and is not such a useful guide to identification although is important in some of the species. Often, the lobes of the style branches have characteristic shapes.

The Californicae or Pacific Coast irises

I. amabilis Eastwood. A synonym of *I. macrosiphon*.

I. bracteata Watson. This species is so called because of the short bract-like leaves all the way up the flower stems, usually tinged with purple-red or pinkish. The proper leaves are very thick and tough up to 1cm wide, which is rather broad for this group of irises, and the whole plant is usually 20–30cm in height. The bracts are held close together around the flower stalk and ovary. There are two flowers per stem and these are about 6–7·5cm in diameter and have a short thick tube only 5–10mm long. The colour is creamy or pale to medium yellow, veined with brown or reddish-purple lines, especially on the falls which have a slightly deeper yellow signal area in the centre. The style branches are relatively small compared with the larger, spreading falls.

I. bracteata occurs in pine forests of the Siskyou mountains in southern Oregon in Josephine County, and in adjacent northern California in Del Norte County, at altitudes of about 450–1000 metres.

I. chrysophylla Howell. This is variable in height from near-stemless to about 20cm tall. The leaves are 3–5mm wide and often slightly grey-green. It has bracts which are held close together and usually enclose two flowers which have a long tube varying from 4·5–12cm in length, rather slender and gradually widening to the apex. The flowers are about 6–7cm in diameter and are cream or very pale yellow with deeper yellow veining and they have narrow, acute, segments so that the flowers lack the substantial appearance of some of the other species with wider segments. The style lobes are very long and narrowly pointed. *I. chrysophylla* occurs in pine and fir forests over a wide area of Oregon and just crosses into northern California. It flowers in April, May or June at altitudes ranging from 400–1200 metres.

I. douglasiana Herbert. This is the best-known of the Pacific Coast group irises in Britain since it is very tough and free-flowering. It is also the most coarse-looking and not the most attractive of the Californians, at least in the forms I have grown. It is very variable, about 15–70cm tall, and has broad leaves, usually deep green, stained red at the base, and up to 2cm wide. The two or three flowers are carried in bracts which are either held together or diverge, and the flower stems are branched which makes *I. douglasiana* unique in its group. The flowers are about 7–10cm in diameter with a tube 1·5–2·8cm in length and the colour varies but is usually in the lavender to purple shades, veined darker and with a yellowish central zone on the falls. Sometimes they are nearly white with darker veining and there is also on record a pure white with a yellow signal patch.

I. douglasiana is a widespread species in the wild, usually not far inland, from southern Oregon southward nearly to Los Angeles, over 1000km in all. The range of habitats is quite wide from open sea cliff tops where I have seen it in flower as early as January, to fields and light woods. Normally it flowers in April, May or June. A characteristic feature of the species is its very triangular cross-sectioned capsule with a short blunt beak-like apex.

I. fernaldii R. C. Foster. A slender species about 20–45cm in height with very grey-green leaves about 7–9mm wide and purple-tinged at the base. The two flowers are about 7–8cm in diameter with a tube about 3–6cm long, widened at the apex and sheathed by non-divergent bracts. The colour is a pale primrose yellow with a deeper yellow median line on the falls and sometimes a faint purple tinge or veining. The style lobes are usually about 1·5cm long and rather blunt at the apex. *I. fernaldii* grows in open woods or woodland fringes in several counties to the north and south of San Francisco at altitudes varying from 250–700 metres. It is one of the few of the Californicae which I can claim to grow from my own wild collection and it does well in full sun in a rather sandy soil in my garden. *I. tenuissima* can be very similar in colouring but has much longer, narrower style lobes and *I. macrosiphon*, too, has yellow forms which can however be distinguished by having an enlarged almost bowl-like portion at the apex of the perianth tube.

I. gormanii Piper. A synonym of *I. tenax*.

I. hartwegii Baker. A very variable plant, recognised in four subspecies by Dr Lenz. It is a deciduous species about 10–30cm in height with leaves only 2–6mm wide, and bracts which diverge and are separated from each other by up to 4cm of stem. There is sometimes only one flower but usually two, about 6–8cm in diameter and of a pale creamy yellow or lavender, often veined darker. They have a very short perianth tube, only 5–10mm long which is quite thick. The falls and standards are narrow so the flower lacks substance and it can hardly be described as a showy plant although in the shorter more compact forms it is quite attractive. *I. hartwegii* subsp. *hartwegii* is a quite widespread inland species in the Sierra Nevada range in California where it inhabits *Pinus ponderosa* forests from about 700–2000 metres altitude. In my garden in full sun it does well and is not entirely deciduous.

I. hartwegii subsp. **australis** (Parish) Lenz. This is a robust version of *I. hartwegii* up to 40cm in height with rather larger, but short-tubed flowers of a dark purple or blue-violet. It also has broader bracts. The leaves have a pinkish-tinged base whereas those of subsp. *hartwegii* usually do not. Subsp. *australis* is confined to the San Bernadino and San Gabriel mountains in southern California at about 1650–2300 metres in pine woods. It is similar in appearance to *I. tenax* but can be recognised by its tough grey-green leaves, those of *I. tenax* being softer-textured and light green, not glaucous. It can be yellow or purple whereas *I. hartwegii* subsp. *australis* is restricted to shades of violet or purple.

I. hartwegii subsp. **columbiana** Lenz. This subspecies also has larger flowers than subsp. *australis*, but as in the preceding two subspecies, the tube is also short (8mm). Its flower colour is restricted to pale yellow with darker yellow veining. The leaves are bright green and wider than those of subsp. *australis*, usually about 1cm. It grows only in Tuolumne County of central California in woodlands at 600–800 metres altitude.

I. hartwegii subsp. **pinetorum** (Eastwood) Lenz. In this variant, which has creamy coloured flowers with deeper yellow veining, the perianth tube is longer, usually about 12–14mm and it is normally narrow in the lower portion and widened in the upper half. The two flowers per stem often open out simultaneously and they have undulate or crisped margins, so presenting a rather frilly appearance. The falls and standards are very narrow so it is not a particularly showy plant in spite of its two flowers opening together. It is restricted to the north end of the Sierra Nevada mountains in California where it grows in pine forest at about 1300–1700 metres.

   I. innominata Henderson. This is a beautiful species and one of the most well-known and popular of the Californian group. It is exceedingly variable in flower colour so that it is impossible to say which form is 'typical' for *I. innominata*. The fact is that one can have a host of different looking plants, all of which are the true species. The features which it possesses are a low habit, usually not more than 15–25cm tall when in flower, very narrow (only 2–4mm) deep green leaves with purplish bases and bracts which are closed around the perianth tube, which is 1·5–3cm in length. The one or two flowers per stem are about 6·5–7·5cm in diameter and have frilly-edged falls and standards and broad style-lobes which, although more or less rounded, are toothed at the margin. In colour the species is at its most variable and it can be found in cream or yellow to orange shades or lilac-pink through pale blue-purples to quite a deep purple. Often the falls are beautifully veined darker, sometimes the standards also, but usually to a lesser degree.

   I. innominata occurs only in south-western Oregon and adjacent north-west California in rather open sunny places, or semi-shade, in humus-rich acid soils. It is often high enough in the mountains to be covered by snow in winter although I have seen specimens from as low as 230 metres. In Britain it is a very satisfactory garden plant, especially for rock gardens and open peat-banks, normally flowering in June.

I. landsdaleana Eastwood. A synonym of *I. purdyi*.

I. macrosiphon Torrey. (Syn. *I. amabilis*). As its name implies, the perianth tube of this species is large; in fact, although it can be as much as 8·5cm in length, it can also be as short as 3·5cm, such is the variability of this iris. It is

SUBGENUS LIMNIRIS, SERIES CALIFORNICAE: *I. innominata*

usually about 15–25cm in height, although there are also very compact forms which are nearly stemless. The leaves, which are considerably longer than the flower stems are grey-green with no pinkish tinge at their bases. The two flowers per stem are long-tubed and about 5–6cm in diameter, enclosed within bracts which are held together, not divergent at the tips. Although the tube varies a lot in length, it has a bowl-like enlargement at the apex which is characteristic. Like *I. innominata*, this species is extremely variable in flower colour from white and cream through yellows of varying shades and pale lavender through to purples and deep violet, usually with delicate veining, especially on the falls. In the centre of the falls there is often a whitish signal area and in one form I have seen at Kew there is also a margin of white to the falls and a yellow line down the centre. This particular form was slightly scented, and Dr Lenz notes that the flowers can be 'delightfully fragrant'. *I. macrosiphon* occurs over a wide area of California, occurring at altitudes of 30–1000 metres on open grassy hillsides or light woods. I have found it distinctly tender in my Surrey garden, but since the species covers such a wide altitudinal range it seems probable that mine was a form from low down and that specimens from higher regions would be frost-hardy.

I. munzii R. C. Foster. This is one of the largest-flowered of the Pacific Coast irises, unfortunately too tender to be grown successfully in countries where there are frosty winters since it occurs in an area near where citrus fruits are grown. It is a robust plant up to 75cm in height with the leaves 1·5–2cm wide, grey-green and with no pinkish stain at their bases. The bracts are divergent and the two are often widely separated by as much as 6–19cm of clear stem. The two to four flowers are 6–7·5cm in diameter and have a thick tube only 7–10mm long. The colour is always in the blue-purple range from pale blue or lavender to deep reddish purple, often veined darker; Dr Lenz has found that in gardens *I. munzii* displays some of the bluest colours known in the rhizomatous irises. Both falls and standards are frequently undulate or even frilly at the margins. It is a native of the southern Sierra Nevada mountains in eastern Tulare County, California, growing in semi-shade, often near streams at about 450 metres altitude. It is unfortunate that we cannot grow this splendid iris outside in Britain although it is certainly worthy of a place in the cool greenhouse where one grows *I. wattii* and *I. formosana*.

I. purdyi Eastwood. A very distinctive iris about 20–35cm in height when in flower with short bract-like leaves all the way up the stems. These and the true leaves are shiny dark green, strongly tinged with pink or red at their bases. The two flowers are about 8cm in diameter, rather flattish because of the spreading standards as well as the falls, and are cream or white, often tinged with lavender, and veined and spotted with purple-pink on the falls. The perianth tube is about 3–5cm long. *I. purdyi* occurs wild in the northern half of California but is not a common plant in its true form, having hybridized with several other species. It grows in slight shade in

SUBGENUS LIMNIRIS, SERIES CALIFORNICAE: *I. tenax*

Redwood forests or other conifer woods in mild areas with a damp climate not far inland from the Pacific coast.

I. tenax Douglas ex Lindley. (Syn. *I. gormanii*). A slender, rather graceful species well known and very successful in British gardens. It is deciduous and grows about 15–35cm in height and has green leaves, stained pink at the base, only 3–5mm wide. The one or two flowers have a tube 6–10mm long, a diameter of 7–9cm and are variable in colour from pale lavender to deep purple-blue and from white to cream and yellow. The bracts are narrow and diverge rather than sheathe the ovary and tube. It occurs over a wide area of south-western Washington State and western Oregon in the open or in slight shade.

In Britain it is very easily grown and has made a good clump in semi-shade in my garden, growing with Hepaticas and Snowdrops. The form I have is a very rich purple-blue, sent to me as a dark form by Marshall Olbrich. The yellow form may be found in literature under the name *I. gormanii* but Dr Lenz regards it as one of the many colour forms of *I. tenax*.

I. tenax subsp. ***klamathensis*** Lenz. This is a variant of *I. tenax* which has a longer perianth tube (1·1–2cm), longer narrow style crests and broader spathes. The flower colour is restricted to buff-yellow or creamy-pinkish usually with brownish or maroon veining. It occurs to the south of the area of subsp. *tenax*, in Humboldt County of north-west California and is a rare plant.

I. tenuissima Dykes. (Syn. *I. humboldtiana, I. citrina*). A slender species usually 15–30cm tall with grey-green leaves only 4–6mm wide and bracts which are held close together around the tube. The two flowers have horizontally spreading narrow falls giving them a rather starry appearance although they are quite large, about 6–8cm in diameter, with a long slender tube about 3–6cm long, the upper quarter or third of which is expanded and nearly cylindrical. The colour is creamy with purple or brown veining on the falls. A distinctive feature of the species is the long, narrow, pointed style lobes which are strongly reflexed. *I. tenuissima* is a native of northern California where it grows in dryish light woodland. It is successful in Surrey, growing well in an open sunny situation with other members of the Californicae section, although one plant rotted off completely during the dull wet summer of 1980.

I. tenuissima subsp. ***purdyiformis*** (R. C. Foster) Lenz. This differs most obviously from subsp. *tenuissima* in having several rather tubby bract-like leaves hugging the stem, and bracts which are similar but even more inflated. Although subsp. *tenuissima* has one to three stem leaves they are not at all inflated and their tips stand well out from the stem, so they appear more leaf-like. The bracts are also longer and straighter, not markedly inflated. In subsp. *purdyiformis* the flower colour is cream or pale yellow with less veining than in subsp. *tenuissima*. It is a plant of shady places beneath pines in the northern Sierra Nevada mountains, in northern California.

In addition to the species and subspecies described above there are a great many natural hybrids and the reader is recommended to consult the most interesting paper by Dr Lee Lens on the 'Hybridization and Speciation in the Pacific Coast Irises' in *Aliso* **4**, pages 237–309 (1959). It appears that the felling of forests in the past has been responsible for the expansion of the territory of some species into the range of others so that hybridization has become possible.

2B(g)   Subgenus Limniris, Section Limniris, Series Longipetalae

A group of only two species (some authorities regard even these to be the same), they are not easily separated from some of the other series on their characters although they do appear to constitute a distinct unit if ecology, geography and morphology are combined. Recent thorough work in the United States by Homer Metcalf suggests that there is only one species in this group. Possibly some formal cultivar names should be given to the variants described below since they do seem to be rather distinct as garden plants. They are robust plants with a west-central-to-western American distribution and are confined to calcareous soils which are very wet in winter and spring and dry in summer. They have tough, widely-spreading, thick rhizomes clothed with the old leaf remains. The fruiting stems persist for a year or more and they carry six-ribbed capsules which taper at both ends.

The Californicae Series is geographically related but here the species occur predominantly in areas which are not excessively wet in spring, in coniferous woods on acid or neutral soils, and are confined to Washington, Oregon and California. They are mostly of much more slender growth and are in fact considered by those American botanists who have made a thorough study of both groups to be rather unrelated to the Longipetalae.

I will follow the American Iris Society's publication *The World of Irises* and use the attractive name 'Rocky Mountain Irises' for the Longipetalae and 'Pacific Coast Irises' for the Californicae Series.

The Longipetalae or Rocky Mountain irises

I. longipetala Herbert. This is a stocky plant with thick stems 30–60cm in height. The dark green, slightly glaucous basal leaves are about 6–9mm wide and usually are just exceeded by or equal the flower stems in length. There are in addition one or two smaller stem leaves. The spathes contain three to eight flowers and one of the striking features of the plant is the large head-like inflorescence on an unbranched stem. Each flower is about 6–7cm in diameter and is veined lilac-purple on a pale or near-white ground, the standards usually much less veined than the falls, which may have a slightly yellowish signal area in the centre. The perianth tube is 5–13mm long and funnel-shaped. *I. longipetala*, as its name says, does have long petals, the large falls being deflexed and having an attractive floppy appearance. They are, however, not distinctly larger than those of the next species and it is probable that the plant should be regarded as a vigorous maritime form of *I.*

missouriensis. I. longipetala grows on open moist hillsides in coastal grassland or mixed deciduous woods in California from around San Francisco south to Monterey. It flowers in March or April, after growing in the winter and spring, and then receives a dry warm summer. *I. longipetala* does reasonably well in sheltered gardens, planted in soils which dry out somewhat in summer but with plenty of moisture in spring. It is recommended that transplanting should be carried out in late summer or early autumn. White forms are recorded but I have not had the fortune to grow one of these.

I. missouriensis Nuttall. (Syn. *I. tolmeiana*). This is rather similar to the description given for that of *I. longipetala*, but it is much more variable, being an extremely widespread iris in the wild. It is a more slender-looking plant with leaves often only 3–7mm wide, usually overtopping the flowers which normally number only 2–3 in each set of spathes. The flowers, which are carried on pedicels up to 20cm long, are very much the same shape as those of *I. longipetala* but range in ground colour from pale blue or lilac to a deeper blue or lavender. The falls are prominently veined, especially towards the base, and there is often a yellowish signal patch. Compared to the large deflexed falls and oblanceolate, erect or oblique standards, the style branches are quite small and insignificant. It has a very wide distribution in western North America from British Columbia southwards to Mexico (but possibly not now represented there) and eastwards into Arizona, South Dakota and Alberta. The habitat is in meadows, on stream banks, in scrub or pine forest but always in areas which are wet until flowering time and dry later. It occurs from low altitudes to 3000 metres and flowers, depending largely upon altitude, between May and July.

The variability of *I. missouriensis* has led to a proliferation of names and it is a matter for relief that the plants attached to these are now considered unworthy of recognition as separate entities. Names which may be encountered are: var. *arizonica*, which has longer, broader leaves (up to 75cm long and 1–1·2cm wide) which are not glaucous and shorter pedicels, less than 10cm long, and branched flower stems; var. *pelogonus*, a shorter version with thick rigid leaves, pedicels less than 10cm long and a relatively short ovary (below 1·5cm); *I. montana* which is much the same as var. *pelogonus*, being a small plant with the deepest flower colouring in the range. Forma *angustispatha* has, obviously, narrow spathes or bracts, which diverge.

In Britain *I. missouriensis* is not a difficult plant to grow, in a sunny position which is well watered in spring. There are various recommendations about the best time to transplant, some considering that spring is best and others the autumn. Certainly summer should be avoided and, in colder districts, the winter also. Since in the wild both this and *I. longipetala* appear to be in most active growth in autumn and early spring it would seem to be open to choice as to when to move or divide the clumps. If autumn is chosen I would suggest fairly early in September to give the plants time to re-establish before the winter.

2B(h) Subgenus Limniris, Section Limniris, Series Laevigatae

These are species of watersides, ditches and swampy grassland in Europe, Asia and North America, and include such familiar plants as the British *I. pseudacorus* and the Japanese Water irises, *I. ensata (I. kaempferi)* and *I. laevigata*. Their distinctive features as a series are mainly that they are tall vigorous plants of wet places, with rather wide leaves and stout rhizomes. The stigma flap on each style branch is bilobed, not a triangular tongue as in the Sibiricae, and the capsules have three ribs giving a roughly three-cornered or nearly cylindrical section, unlike the six-ribbed capsules of the Hexagonae and Spuriae. Being water plants they produce seeds which are usually capable of floating, and the capsules are thin-walled, breaking up irregularly or rotting away rather than splitting open to release their seeds as in most iris species.

Cultivation

All the species in the Series Laevigatae are easily grown in wet places in the garden and make admirable plants for the margins of ponds and streams. The soil should not only be supplied with plenty of moisture but also needs to be reasonably rich in humus such as old rotted manure or leaf mould. Some of them are also capable of being grown in open borders but do need a very good supply of water through the summer months, and a lot of water-retentive peat or leafmould incorporated in the soil. Propagation is by seed or by division of the clumps of rhizomes, in September preferably, but also possible in April.

The Laevigatae irises

I. ensata Thunb. (Syn. *I. kaempferi*). This is a frequently-grown plant in gardens, available in many different cultivars. The wild form grows about 60–90cm in height and has the leaves shorter than the flower stems, about 20–60cm in length, and about 4–12mm wide. They have a very prominent midrib unlike those of *I. laevigata* which are quite smooth. The flower stems are either unbranched or occasionally have one branch, and carry three or four flowers which in wild plants are purple or reddish-purple and about 8–15cm in diameter. The falls have a yellowish haft, the yellow spreading to the base of the elliptical or ovate blade. The standards are smaller and erect and the perianth tube is about 1–2cm in length. It is said that the anthers are yellow in *I. ensata* and white in *I. laevigata* but I cannot confirm or contradict this. *I. ensata* occurs wild in Japan, northern China and the eastern USSR, in the Sakhalin, Amur and Ussuri regions, in wet grassy places up to about 2400 metres altitude and flowering in June or July. Although often confused with *I. laevigata* in the wild form it is a much more slender plant and can be immediately recognized by the prominent midrib. There are many garden forms with larger flowers, sometimes double, and often with spreading, rather than erect, standards.

I. kaempferi Siebold. A synonym of *I. ensata*

I. laevigata Fischer. This is similar in stature to *I. ensata* but is a stouter plant with leaves 1·5–4cm wide, lacking the prominent midrib of the latter. The flower stems often have one branch and are usually two-to-four flowered with green sharply keeled spathes. The flowers in wild form are blue-purple or white, about 8–10cm in diameter with a tube up to 2cm in length. There is a narrow, yellow haft to the falls, extending up to the obovate or ovate-elliptic blade. The standards are oblanceolate, much smaller than the falls, and erect. Unlike *I. ensata*, in which the anthers are yellow, it is reported in Japanese Floras that *I. laevigata* has white anthers but I have not had the opportunity to observe living wild specimens of both species.

I. *laevigata* is a widespread species in eastern Asia, from Lake Baikal and the Altai regions of central Russia eastwards through northern China and Korea to Japan. It occurs in swamps and on lake margins and it flowers in June or July. As with *I. ensata* there are many named garden forms.

I. pseudacorus Linn. This is the familiar Yellow Flag iris, the only yellow species in its series and such a distinctive plant that it cannot be confused with any other. It grows about 75–160cm in height and has greyish-green leaves 1–3cm wide. The four to twelve yellow flowers are about 7–10cm in diameter and they usually have some brown or violet veining on the falls, and a darker yellow zone. Some forms, however, lack this dark blotch, as in var. *bastardii*. The standards are only 2–3cm long and are narrowly oblong.

I. *pseudacorus* (i.e. 'false acorus') is so called because its foliage is thought to resemble that of the Sweet Flag, *Acorus calamus*. It is a common waterside plant in Europe including Britain and extends to western Siberia, the Caucasus, Iran, Turkey and North Africa. It flowers in June, July or August in and alongside lakes, ponds, streams and ditches, often covering large areas. It is also naturalized in several other parts of the world. Medicinally the plant has been used as a strong purgative, and the herbalist Meyrick gave full warning of its efficacy. He noted that it 'has been found to procure plentiful evacuations from the bowels when all other means have proved ineffectual.'

I. shrevei Small. A synonym of *I. virginica* var. *shrevei*.

I. versicolor Linn. This robust clump-forming plant has stout creeping rhizomes giving rise to erect or arching leaves about 1–2cm wide and stems 20–80cm in height, equalling or slightly exceeding the leaves. The branching flower stems carry several flowers, each about 6–8cm in diameter and usually some shade of violet, blue-purple, reddish-purple, lavender or dull slatey-purple. The falls are widely spreading and often have a greenish-yellow blotch at the centre of the ovate blade, surrounded by a white area variegated with purple veins, this continuing down the haft. A white form is also known, and one particularly good reddish-purple variant is often grown under the name of var. *kermesina*. The standards are erect, much smaller and usually a little paler than the falls, with veining on the haft. *I. versicolor* is a very widespread plant in eastern North America from eastern Canada southwards to Texas. It grows in marshes, swamps, wet meadows and on lake shores and flowers in May, June or July.

As a garden plant *I. versicolor* presents no problems and may be grown either in the water garden or in a herbaceous border if plenty of humus is incorporated in the soil. The rhizomes and roots have been used medicinally in the past in parts of America in the treatment of diseases of the liver and of dropsy.

I. virginica Linn. The Southern Blue Flag. This is sometimes considered by botanists to be inseparable from *I. versicolor*. I am not familiar with either in the wild and would not care to judge. Currently it is treated as a separate species by several American Floras. As its common name suggests, it has bluer flowers and is confined to a more southerly part of the United States. The height varies, 30–100cm, and the stems are often arching, falling to the ground in the fruiting stage. The leaves, 1–3cm wide, are soft and flopping over at the tips. In typical *I. virginica* there may be one short branch on the stem but usually it is simple. The one to four flowers are 6–8cm in diameter with spreading falls of blue, violet, lilac, lavender or occasionally pinkish-lavender. In the centre of the 3–4cm wide, oblong or obovate blade there is a prominent yellow hairy patch which helps to distinguish *I. virginica* from *I. versicolor*. The standards are erect and smaller, usually narrowly obovate or spathulate in shape. An albino form is recorded but I have not seen this in cultivation. *I. virginica* grows in marshes, damp pinewoods, ditches and wet grassy places in Florida and eastern Texas northwards to south-eastern Virginia. It flowers from May to July. In cultivation in Britain it may be treated in much the same way as *I. versicolor*. Although very similar in appearance to the latter it may be recognized by the flower colour which is usually in the bluer end of the spectrum (reddish-purple in *I. versicolor*) and by the yellow hairy patch on the falls. It does, of course, belong to the non-bearded group of irises, and this pubescent patch in no way resembles the beard of a 'pogon' iris.

I. virginica var. **shrevei** (Small) Anders. (Syn. *I. shrevei*). This is a variant in which the inflorescence has long branches (simple or with one short branch in var. *virginica*) and capsules which are twice as long as wide, usually 7–11cm long. In var. *virginica* they are only 4–7cm long and are much less than twice as long as broad.

2B(i)   Subgenus Limniris, Section Limniris, Series Hexagonae

The Hexagonae group consists of five species of large-flowered water irises, so called because of the six-ribbed ovary and capsules of *I. hexagona*, although this feature is not so clear in some of the other species. The seeds are very large with a cork-like coat which enables them to float, this frequently being the method of seed dispersal in waterside plants. These irises inhabit marshes and swamps in the southern United States from Florida to Texas and northwards through the Mississippi valley to the Ohio river. The Hexagonae, or 'Louisiana Iris' as they are often called, are robust plants with branched infloresences, leaf-like bracts and large flowers. There are presently considered to be five species and a great many natural and man-made hybrids, but these are of no great horticultural importance in Britain, being either tender or not very free-flowering. Anyone interested in finding out more about the development of this group as garden plants is recommended to read the appropriate chapter in *The World of Irises* published by the American Iris Society in 1978.

In 1931 a great many 'species' were described by J. K. Small and E. J. Alexander *(Contributions from the New York Botanical Garden,* **327**). However these have since been interpreted as belonging to a small group of related species and their natural hybrids. *I. nelsonii* is also regarded as a natural hybrid but is now stabilized and occupies a separate ecological niche, so is usually treated taxonomically as a distinct species. The rest of the numerous specific names which were attached to these hybrids and variants are not listed here.

Cultivation

In British gardens this is not a highly successful group, although *I. fulva* and *I. brevicaulis* do reasonably well in the south. They need a humus-rich soil, either acid or neutral, which is permanently moist through the growing season. Since they rapidly exhaust the soil they need replanting every second or third year in freshly prepared beds. Rapid and robust growth is required in spring in order to get a reasonable amount of flower and it is recommended that liquid manure is given as soon as the weather warms up and growth begins. *I. giganticaerulea* and *I. hexagona* are reputed to be tender although I cannot write from personal experience about this. Mrs E. Osborn, writing in the British Iris Society Species Group's Notes, regards the Louisiana irises as 'worthwhile and trouble-free garden plants' so it is possible that I should try again with more adequate preparation of the site, for I must admit to having tried only *I. fulva* in a bed of mixed plants. Her recommendations are basically as given above, but cow manure and compost are added to the soil before planting, and a mulch of compost in late autumn. Division of the clumps is carried out in September.

In North America, Hexagonae irises are so popular that there is a Society for Louisiana Irises, formed in 1941 and dedicated not only to their cultivation and development but also to the preservation of the species in the wild. The alteration of the habitat—for example by turning swamps into rice fields or by draining the land for farming—seems to be the main threat to their continued existence.

In the United States they are recommended not only as bog plants but also for the herbaceous border and as cut flowers, the succession of blooms apparently lasting for a considerable period.

The Hexagonae or Louisiana irises

I. brevicaulis Rafinesque. (Syn. *I. foliosa; I. lamancei*). A rather short plant for this group, from about 30–50cm in height, it has long slender rhizomes and stems which zigzag at the nodes. The 3–6 large stem leaves overtop the flowers and are about 2–5cm wide, giving the plant a very leafy effect. There are several flower heads, produced both terminally and from the axils of the leaves, and these give rise to gorgeous bright blue-violet flowers about 6·5–10cm in diameter. They have the large broadly ovate falls reflexed and the much smaller standards spreading out widely, rather than erect or drooping. In the centre of the falls is a yellow median band, and the haft is usually veined whitish-green.

It is a native of the Mississippi river basin in damp pastureland, flowering in March or April.

I. foliosa Mackenzie & Bush. A synonym of *I. brevicaulis*.

I. fulva Ker-Gawler (Syn. *I. cuprea*) A robust plant 45–80cm in height with slender green rhizomes producing fairly straight or only slightly zigzag stems and with leaves 1·5–2·5cm wide, the tips of which are drooping. The flowers are smaller than those of the other species in the group, about 5·5–6·5cm in diameter, but are striking for their red, coppery-red or orange-red colours with both falls and standards flopping downwards, leaving the small style branches standing up at an oblique angle. It grows wild in wet, sometimes partially shady, meadows and stream banks in the Mississippi valley and usually flowers in April or May.

I. fulva is one of the easiest of the species and is a very satisfactory plant in a swampy place adjacent to a stream or pool. It seems to grow very well in the sort of conditions where trilliums and the candelabra primulas thrive. A yellow form and an albino are known, but I have never had the fortune to see the latter in cultivation.

I. × *fulvala* is sometimes grown in gardens. It is a purple-red flowered hybrid, raised by W. R. Dykes in 1910, between *I. fulva* and *I. brevicaulis* (= *I. lamancei*).

I. giganticaerulea Small. This is taller than *I. fulva*, usually 70–180cm, and has thick rhizomes and straight, rigid stems which carry long bluish-green stem leaves. The flowers are very large, up to 14cm in diameter but not floppy in appearance. The blades of the falls spread out nearly horizontally or are only slightly deflexed and the standards are erect. Normally the colour varies from clear pale blue to deep indigo with a yellow signal ridge in the centre of the falls.

It occurs in the Gulf of Mexico region of Louisiana, Texas and Mississippi in coastal marshes rich in humus. Flowering time is in March or April and it is regarded as a useful plant in American gardens for it is generally earlier than the other species.

It is not known in British gardens as far as I can ascertain, being undoubtedly tender.

I. hexagona Walter. This occurs farther to the east than the others and is not primarily a 'Louisiana' iris in distribution although it does of course belong to the same Series as the rest of the species. It has not been used in the development of the great range of Louisiana hybrids, these being crosses between the other four. *I. hexagona* has branching stems and grows about 30–90cm in height with long stem leaves about 2·5cm wide. The large flowers are 10–12cm in diameter and are some shade of bluish-purple or lavender with a yellow signal area in the centre of the falls. The haft is often speckled white or yellowish, shading to green on the lower part. The standards are upright and much narrower and more pointed than the deflexed falls. Albino forms are also known to exist.

I. hexagona flowers between March and May and inhabits swampy areas in the south-eastern United States, in South Carolina, Florida, Alabama, Georgia and Louisiana.

I. lamancei (Gerard) Lynch. A synonym of *I. brevicaulis*.

I. nelsonii Randolph. This has stems 70–110cm tall bearing several short stem leaves and there are usually a few branches. The basal leaves are 1–3cm wide, frequently with drooping tips, and are of a pale green. As in *I. fulva* the large, up to 10cm-diameter, flowers have reflexed falls and standards although these are much larger than in *I. fulva* and the colour is reddish-purple. Yellow forms are also known.

This iris is considered, with some certainty, to have arisen by hybrization between *I. fulva* and *I. giganticaerulea*, but is now a stable entity and worthy of recognition as a species. Before their descriptions as *I. nelsonii* the populations of these reddish-flowered irises were known as the 'Abbeville Reds' since they are restricted in the wild to the Abbeville swamp of Vermilion Parish in central-south Louisiana.

2B(j)   Subgenus Limniris, Section Limniris, Series Prismaticae

A series containing one species only, *I. prismatica*, which is a native of North America. The features which distinguish it from other series are the thin widely-creeping rhizomes, the tall slender stems with smallish brown bracts from which the flowers are exserted on long pedicels, and the three-cornered capsule with one rib at each of the wing-like corners. The seeds are smooth-coated.

I. prismatica is an easily-grown plant in Britain, given the right conditions. I find that it thrives and makes large clumps, planted in sandy-peaty soil in the partial shade of some fir trees. Here it receives plenty of moisture in winter and spring but tends to dry out in summer when the fir tree roots are most active.

I. prismatica Pursh ex Ker-Gawl. A slender grassy plant with thin far-creeping rhizomes which send up tufts of erect, glaucous leaves only 2–7mm wide and 50–70cm in length. The flower stems reach 30–80cm in height and are thin and wiry, usually not straight, and often have one branch, this carrying only one flower whereas the terminal head has two or three. The flowers are about 5·5–7cm in diameter and are carried on pedicels up to 4cm long, held well above the smallish brown papery bracts. There is a short tube only 2–3mm long and widely spreading, rather arching falls which have a small ovate blade and a long narrow haft. The colour is predominately pale violet or violet-blue, but the haft of the falls is violet-veined on a greenish-white ground colour. The standards are oblanceolate, suberect and violet as are the style branches which have smallish, recurved lobes. After flowering the pedicels elongate to 7–10cm in length and carry 3–5cm long capsules which are sharply three-cornered to the extent of having almost wing-like projections.

I. prismatica is a native of the Atlantic coastal States of North America from Nova Scotia southwards through to North and South Carolina. It also occurs away from the coast, in the southern Appalachians of Georgia and Tennessee, where it is perhaps slightly more robust and is given the name of var. *austrina*. It inhabits marshes, swamps and damp fields near the sea and flowers in June or July. In Britain it usually flowers in June and although the rather 'spidery' flowers are not showy it is a delicate species with a quiet charm which is lacking in many of the larger more gaudy species. I am fortunate in having a beautiful white form, given to me by Roy Davidson some years ago. The flowers are a creamy-white and shade to green on the haft of the falls. It grows very vigorously in a semi-shady peat bed.

2B(k) Subgenus Limniris, Section Limniris, Series Spuriae

I. spuria, the typical species of this series, often grows in damp saline soils and it is tempting to give the group the common name of 'Salt Marsh Irises'. This is not however the habitat in which one finds all the species and some of them are subalpine plants growing in turf, while others occur in dry scrubland or steppe conditions. They do however all have several morphological features which enable them to be grouped together.

The shape and character of the capsules and seeds is probably the most important: the capsule has a narrow beak at the apex and there are six ribs running along it longitudinally, these being in three pairs rather than equally spaced around the capsule. The seeds are hard but they are enclosed within a loose papery coat. Other characters which one can add are the possession of a two-pronged stigma, and the curious feature of having three drops of nectar on the outside of the perianth tube at the base of the segments. This is known to attract ants but its true purpose is not known for it seems hardly beneficial for pollination purposes, since insects need not go near the anthers in order to obtain the nectar.

In general growth habit the Spurias are mostly tough plants having rather woody rhizomes with wiry roots and producing leathery leaves. The flower stems are either unbranched or if branched then these are held close to the main stem giving a spike-like effect. The flowers are fairly typically iris-shaped with no special peculiarities. In most species the falls have a slightly winged haft and a constriction between the haft and the ovate, elliptical or orbicular blade. There is of course no proper beard although in some species there is a slightly pubescent area in the lower and central part of the falls. This pubescence consists of unicellular papillae only, not multicellular hairs.

The notes I have given about the species in this rather complex series are based where possible on personal experience or on observations of herbarium material, but I have also had to rely on published information. The following sources have been very useful: 'Wild Iris species of the USSR' by Dr G. I. Rodionenko in *The Iris Year Book*, 1964; Dr Lee W. Lenz in *Aliso*, **5**, number 3 (1963) and Ben R. Hager in *The World of Irises*, published by the American Iris Society (1978).

Cultivation

The smaller species such as *I. pontica*, *I. sintenisii* and *I. kernerana* are good rock-garden plants, growing well in full sun or semi-shade with a well-drained humus-rich soil which does not dry out excessively in summer. The taller species require only good soil in full sun and do not necessarily require wet or saline conditions such as they often prefer in the wild. They make good herbaceous border plants and respond well to plenty of well-rotted compost incorporated before planting. On the whole they do not seem to like sandy soils in the east of England, preferring the moister, milder climate of the west.

Propagation can be by division of clumps in autumn or early spring but care must be taken with the newly planted rhizomes to make sure that they do not dry out. It is undoubtedly better if possible to grow new plants from seed and to plant out the seedlings into their permanent positions as soon as possible to avoid disturbance.

The Spuria iris species

I. albida Davidoff. A synonym of *I. ochroleuca*.

I. aurea Lindley. A synonym of *I. crocea*.

I. brandzae Prodan. A synonym of *I. sintenisii* subsp. *brandzae*.

I. carthaliniae Fomin. A synonym of *I. spuria* subsp. *carthaliniae*.

I. colchica Kem.-Nat. The type specimen of this, from the Caucasus (Georgia), looks identical to *I. graminea* and I think that the name must be regarded as a synonym, although I must admit to not having studied living material from Caucasia.

I. crocea Jacq. (Syn. *I. aurea* Lindl.) This is said to have originated in Kashmir and today it is certainly still growing in the Valley of Kashmir at 1600–2000 metres. It is, however, associated with cemeteries and it is not clear whether it is a truly wild plant in this area, or an introduction from places unknown. It is a vigorous plant, up to 1·5 metres in height with sword-shaped leaves to 75cm long and 1·5–2cm wide. The inflorescence consists of a terminal group of

SUBGENUS IRIS, SECTION IRIS

1   *I. pallida* subsp. *cengialtii*

2   *I. imbricata* in the Elburz
      Mountains, Iran

3   *I. attica* in Turkey near Bâlikesir

SUBGENUS IRIS, SECTION ONCOCYCLUS

4 *I. meda* in Iran near Miyaneh

5 *I. paradoxa* in Turkey near Van

SUBGENUS IRIS, SECTION ONCOCYCLUS

6 *I. barnumae* in Turkey near Van

7 *I. atropurpurea* in the Negev desert, Israel

8 *I. iberica* subsp. *lycotis* in western Iran

SUBGENUS IRIS, SECTION REGELIA

9   *I. afghanica*

SUBGENUS IRIS,
SECTION PSEUDOREGELIA

10   *I. korolkowii*

11   *I. kamaonensis*

SUBGENUS LIMNIRIS, SECTION LOPHIRIS
12   *I. cristata*

13   *I. japonica* 'Variegata'

SUBGENUS LIMNIRIS, SERIES CHINENSES
14   *I. minutoaurea*

SUBGENUS LIMNIRIS, SERIES TRIPETALAE
15   *I. setosa* north-east of Anchorage, Alaska

SUBGENUS LIMNIRIS, SERIES CALIFORNICAE
16   *I. innominata*

SUBGENUS LIMNIRIS, SERIES LONGIPETALAE
17   *I. longipetala*

SUBGENUS LIMNIRIS, SERIES LAEVIGATAE
18   *I. laevigata*

19   *I. versicolor*

20   *I. spuria* subsp. *musulmanica* in Turkey, east of Erzincan

21   *I. sintenisii* in Turkey at Abant, near Bolu

SUBGENUS LIMNIRIS, SERIES TENUIFOLIAE
22   *I. songarica* in Iran

SUBGENUS LIMNIRIS, SERIES UNGUICULARES
23   *I. unguicularis* 'Cretensis' near Gythion in southern Greece

SUBGENUS NEPALENSIS
24   *I. decora*

SUBGENUS XIPHIUM
25   *I. serotina*

28   *I. aucheri* in south-east Turkey

29   *I. doabensis*   30   *I. nicolai*

SUBGENUS SCORPIRIS

31   *I. kuschakewiczii*

32   *I. pseudocaucasica* in Iran
near Khoi

33   *I. zaprjagajewii*

SUBGENUS HERMODACTYLOIDES
34   *I. reticulata* in Turkey near Erzurum

35   *I. histrioides*, the wild form from Amasya, Turkey

SUBGENUS HERMODACTYLOIDES

36   *I. histrio* from southern Turkey

37   *I. pamphylica* in Turkey
near Antalya

38   *I. danfordiae*, a wild form
from northern Turkey

flowers and sometimes up to three branches which stay erect and close to the stem. The individual flowers are about 12–18cm in diameter and are deep golden-yellow; the blade of the falls is 4·5–5cm long and 2–2·5cm wide and is wavy at the margins, while the haft is shorter than the blade, usually

SUBGENUS LIMNIRIS, SERIES SPURIAE
1 *kernerana*
2 *I. sintenisii*
3 *I. spuria* subsp. *carthaliniae*
4 *I. graminea* var. *pseudocyperus*

about 3cm long. The standards are erect and oblanceolate, also somewhat crinkled at the edges.

I. crocea flowers in June in Kashmir and is a beautiful and stately garden plant, easily grown in a sunny border.

I. daenensis Kotschy ex Baker. A synonym of *I. spuria* subsp. *musulmanica*.

I. demetrii Akhv. & Mirz. A synonym of *I. spuria* subsp. *demetrii*.

I. desertorum Ker-Gawl. A synonym of *I. spuria* subsp. *halophila*.

I. farreri Dykes. See comments under *I. polysticta*.

I. graminea Linn. (Syn. *I. colchica*). Although a rather leafy plant *I. graminea* is an attractive and useful garden iris with deliciously fruity-scented flowers. It grows about 20–40cm in height and has up to three, 5–15mm wide, long-tapering leaves carried on the flattened two-angled stem, the upper one of which is overtopping the flower considerably. The bracts are unequal, the lower one being the larger and somewhat leaf like, and they enclose one or two flowers which are about 7–8cm in diameter. They have purple standards and style branches and the falls are violet at the tip of the blade, with violet veins on a white ground in the centre. The prominently winged haft is often greenish or brownish tinted, as is the lower part of each style branch. The capsules are ellipsoid, about 2·5–4cm long with six obvious longitudinal ridges.

I. graminea is widespread through central Europe from north-eastern Spain to western Russia, and is also present in the north and western Caucasus. It grows in semi-shade, and the best plants I have grown are on a raised peat bed in humus-rich soil where they flower freely in May or June.

I. graminea var. **achtaroffii** Prodan. This is a variant in which the flowers are said to be yellowish-white.

I. graminea var. **pseudocyperus** (Schur) Beck. A larger and more robust plant than var. *graminea*, and the flowers are unscented. Although it is always stated to be from Roumania and Czechoslovakia I have a similar plant, given to me by Richard Gorer, which was collected in Spain.

I. graminifolia Freyn. A synonym of *I. kernerana*.

I. gueldenstadtiana Lepechin. A synonym of *I. spuria* subsp. *halophila*.

I. halophila Pallas. A synonym of *I. spuria* subsp. *halophila*.

I. humilis M. Bieb. A synonym of *I. pontica*.

I. kernerana Aschers. & Sint. This delightful slender species is one of the most attractive of the smaller Spurias, suitable for rock gardens or the front of mixed borders. It is about 20–30cm in height, rarely up to 55cm, and has linear leaves up to 5mm wide (occasionally to 1cm in very vigorous specimens). These are usually a little shorter than the flowering stems. The bracts enclose two to four flowers, are rather inflated and become straw-like in texture towards the fruiting stage. Each flower is about 7–10cm in diameter and has arched-recurved falls, the blade curling right over so that its tip sometimes touches the stem. The style branches also arch over, following the curve of the falls, but the standards remain erect. Basically the colour is a soft lemon yellow or deep cream with a deep yellow blotch in the centre of the blade of the falls, becoming paler towards the margins. There is usually a faint scent but it is not strong enough to be noticeable in the garden. The individual flowers are not long-lived and rapidly wilt but remain attached, detracting somewhat from the otherwise rather 'clean-cut' appearance of this graceful plant. The elliptical blade of the falls is about 1·5–2cm wide, narrowed abruptly to the 5mm wide haft. In fruit *I. kernerana* has tubby capsules 2–3cm long and 1–1·5cm wide with a narrow beak up to 1cm long.

This is a Turkish endemic, confined to the northern half of the country from Balikeşir province in the west to Erzincan in the east. The habitat is usually in dryish oak scrub, open pine woodlands or in sparse grassland at altitudes of 300–2300 metres and the flowering time is May or June.

I. kernerana grows well in Surrey in an open sunny position in a sandy loam which is neutral or slightly acid. It needs plenty of moisture in the spring but cannot tolerate waterlogged conditions, so good drainage is also necessary.

I. klattii Kem.-Nat. A synonym of *I. spuria* subsp. *musulmanica*.

I. lilacina Borbas. This is a synonym of *I. spuria*, probably subsp. *sogdiana*.

I. longipedicellata Czecz. This was described from Eldiven Dağ in Turkey and appears to be identical to *I. orientalis*.

I. lorea Janka. Described from Italy, this plant closely resembles *I. sintenisii* and should probably be regarded as synonymous with it. It is said to have long, rather pale green leaves and green bracts.

I. ludwigii Maxim. A rare species which I have not seen, it is probably not in cultivation. I have extracted information from Russian literature for the following notes. It has a flower stem only 2–3cm long, but the narrow (5mm wide) leaves can reach 25–40cm, greatly overtopping the flowers as in the better-known *I. pontica*. There are two flowers per stem, produced in succession from three bracts, and they are violet-blue. Its characteristic feature seems to be that the distance between the top of the ovary and the top of the perianth tube is about equal to the length of the segments. I hesitate to call this portion the perianth tube since it seems likely that, in the description I have seen, the true tube and the beak of the ovary are being combined into one measurement. In other species this portion is only about one third as long as the segments. There is also, according to Dr Rodionenko, a minute longitudinal beard on the falls. However, setting aside these features, the general appearance and stature of the plant distinguishes it from most other Spurias.

I. ludwigii flowers in May in the wild. It is an endemic of the southern Altai region of Russia where it grows on gravelly slopes and in open steppe country.

I. maritima Lam. A synonym of *I. spuria* subsp. *maritima*.

I. marschalliana Bobrov. A synonym of *I. pontica*.

I. monnieri DC. This beautiful Spuria iris is probably of hybrid origin and this matter has been studied and discussed by Dr Lee Lenz in *Aliso*, **5**, number 3 (1963). It has the same stature as *I. orientalis (I. ochroleuca)* but the flowers are wholly lemon yellow in colour. The style lobes are very short and strongly recurved, quite unlike the longer, more narrowly triangular ones of *I. crocea* and *I. orientalis*. The idea that it is a garden hybrid between the predominantly white *I. orientalis* and the deep yellow *I. crocea* might account for the intermediate colour but not the distinctive style crest shape.

There are, it seems to me, two more factors which are important in deciding the probable origin of *I. monnieri*.

Firstly, W. R. Dykes raised seedlings from self-pollinated *I. monnieri* and found that the majority had the appearance of *I. orientalis*. Secondly, the blade of the falls is orbicular in *I. monnieri* and, if *I. crocea* had been used as one of the parents, one would expect the shape to be rather more elongated. Dr Lenz notes that 'first generation hybrids between *I. ochroleuca* [*I. orientalis*] and *I. crocea* have falls somewhat tapered like those of *I. crocea* and quite unlike those of *I. monnieri* as shown in Redouté's painting which accompanied the original description.'

It seems more likely that *I. monnieri* is the result of a cross between *I. orientalis* and a deep yellow Turkish Spuria which is in cultivation under the temporary name, coined by Dr Lenz, of 'Turkey Yellow'. This interesting iris has short recurved style lobes similar to those of *I. monnieri*. It occurs in apparently wild situations in several parts of Turkey and a description can be found in this book in the alphabetical sequence below, under 'Turkey Yellow'. Both *I. orientalis* and 'Turkey Yellow' occur wild on the Turkish mainland. The former is recorded in the east Aegean Islands and the latter has been collected in Muğla province of Turkey, which is very close to the island of Rhodes. The plant described by De Candolle in 1808 as *I. monnieri* was growing in a garden at Versailles where it was known as the 'Iris de Rhodes'. It is not beyond the realms of possibility therefore that *I. monnieri* is a natural hybrid introduced originally from the eastern Aegean region.

I. musulmanica Fomin. A synonym of *I. spuria* subsp. *musulmanica*.

I. notha M. Bieb. A synonym of *I. spuria* subsp. *notha*.

I. ochroleuca Linn. A synonym of *I. orientalis* Miller. (Syn. *I. albida*).

I. orientalis Miller (Syn. *I. ochroleuca*). This is the middle-eastern Spuria iris which is best known as *I. ochroleuca* but this name is unfortunately later and therefore unacceptable under the International Rules of Nomenclature. It is a robust plant growing about 40–90cm in height with the basal leaves 1–2cm wide. There is usually one branch to the inflorescence, sometimes more, and it is held close to the main stem. The papery bracts enclose several flowers, opening in succession, and these are about 8–10cm in diameter. It is scarcely variable in colour being almost wholly white except for a large yellow signal area on the orbicular blade of the falls. This blade is up to 3cm in diameter and is abruptly narrowed to a slender haft which is sometimes a little pubescent and is longer than the blade. The erect standards and style branches are white.

Although *I. orientalis* is said to have been collected in Yugoslavia it does not appear to grow there naturally and it must be assumed that this is either an error or refers to a naturalized or cultivated specimen. Its distribution is mainly in Turkey, apparently not occurring farther east than Kayseri in the centre of the country. It is also recorded in the eastern Aegean Islands of Samos and Lesbos, and there is a small outlying area in north-eastern Greece. Like several of the Spuria species it is primarily a plant of saline soils, often seen growing in a narrow band along the edges of irrigation channels or on marshy ground, usually at low altitudes of about 150–1030 metres. In the garden it is an easily grown and free-flowering species requiring only a reasonably sunny position.

I. polysticta Diels. The far-eastern Spurias are mainly known as herbarium specimens only and the details given must therefore be open to a certain amount of doubt, even to the extent of querying whether the species are correctly placed in this series. However, good material of *I. polysticta* exists and it appears that it should be classed with the Spuria group. It grows to about 30cm in height and has tough brown persistent leaf bases at the apex of the rhizome. The leaves are only 2–4mm wide, rather rigid and strongly ribbed, the longest more or less equalling the flower stem in length. The bracts enclosing the flowers are about 8cm long and not markedly inflated as in some of the species. The flowers, 6cm in diameter, appear to be lavender-violet with the widely spreading falls prominently veined on a pale ground and spotted along the centre of the haft. Both the falls and the standards are narrow, at most 8–10mm across the blade, so that the flowers are rather 'spidery' in appearance. As with several other species, the ovary has a narrow beak at the apex while the very short true perianth tube is wider than this and almost cup- or bowl-shaped. The style branches are strongly arched over the haft of the falls.

I. polysticta is described as growing in mountain meadows at 3200 metres in the northern part of Szechuan province in China, and flowering in July.

I. farreri Dykes appears, from the herbarium material available, to be the same as *I. polysticta*. Certainly from dried specimens, I have been unable to find any significant differences. It was seen by Farrer in south-western China, in west Kansu, growing in alpine pastures at about 2700–3700 metres altitude. A feature which is visible on one of the specimens I have seen is the two-toothed stigma, one of the characteristics of the Series Spuriae.

I. pontica Zapal. (Syn. *I. humilis* M. Bieb.; *I. marschalliana*). This is a dwarf tuft-forming species, about 10cm in height when in flower but with overtopping grassy leaves 2–4mm wide and up to 20–40cm long. The solitary flowers have stems only 1–4cm long and they are about 5–6cm in diameter with a nearly orbicular blade to the falls, separated from the winged claw by a definite constriction. The colour is violet with a whitish or pale greenish-yellow, violet-veined area on the haft and the centre of the blade of the falls, and the whole flower is finely darker-veined. The capsule is produced at or only just above ground level. It occurs wild in dryish grassy steppe country in central and north-eastern Roumania, western Ukraine and the Caucasus, possibly extending eastwards into Russian Central Asia although its exact distribution does not seem to be well known.

I. pontica is an excellent hardy little plant for pot cultivation in an alpine house and for the rock garden as well.

I. prilipkoana Kem.-Nat. A synonym of *I. spuria* subsp. *demetrii*.

I. pseudocyperus Schur. A synonym of *I. graminea* var. *pseudocyperus*.

I. sintenisii Janka. This primarily Balkan species is one of the smaller, more slender Spurias, forming large tufts of growth and normally reaching about 10–30cm in height when in flower. The linear leaves are only 2–5mm wide and usually overtop the flowers. I have seen large clumps in Turkey up to 35cm across, densely leafy and well-flowered, but it seldom makes such compact plants in cultivation. The flowers are often solitary and are about 5–6cm in diameter with rather narrow falls and standards, only 8–12mm across the blade. They are deep violet-blue in overall appearance, sometimes with an almost metallic appearance, but the falls are in reality veined violet on a white ground. The bracts are papery and are similar to each other, not as in *I. graminea* where one is larger and leaf-like. The fruit of *I. sintenisii* is a short tubby capsule about 1·5cm long with a long slender beak, also about 1·5cm in length.

It occurs wild in south-east Europe in Yugoslavia, Roumania, south-west Russia, Bulgaria, Albania and Greece and is also in Turkey, mainly in the north-west and the Pontus mountains. It is a plant of dryish scrub or low mountain meadows, sometimes in wood clearings, at about 900–1500 metres altitude and it flowers in June or July.

I. sintenisii subsp. **brandzae** (Prodan) D. A. Webb & Chater. (Syn. *I. brandzae*). I am not familiar with living material of this plant, but it does appear to be distinct and worthy of cultivation. The leaves are only 1·5–3·5mm wide with fewer veins than in subsp. *sintenisii*, and the bracts are strongly inflated. It occurs in north-eastern Roumania in damp saline places. The Roumanian botanist Prodan regarded it as a separate species, *I. brandzae*, distinct from *I. sintenisii*. He described a form of it, forma *topae*, in which the bracts are 6·5–7·5cm long and rather slender.

I. sogdiana Bunge. A synonym of *I. spuria* subsp. *sogdiana*.

I. spathulata Lam. A synonym of *I. spuria*, probably subsp. *spuria*.

I. spuria Linn. This species, or rather species aggregate, has caused much confusion among botanists and will no doubt continue to do so until a thorough study of wild populations over its complete range from Europe to central Asia can be made. Meanwhile I can attempt to simplify the mass of names by combining some of the available published information together with the limited amount of fieldwork I have done on the group in Turkey and Iran. I accept that the divisions are not altogether satisfactory but for those who despair, the name *I. spuria* sensu lato is still available!

It is worth quoting the *Flora USSR* **4** (1935) on the subject of this group. 'I. halophila, together with a number of other kindred species (*I. sogdiana* Bunge, *I. musulmanica* Fomin, *I. violacea* Klatt and *I. notha* M.B.) form a cycle of closely related species that are very difficult to distinguish and are in turn extremely close to the south European *I. spuria* L. and the Persian *I. daenensis* Kotschy. It would perhaps be correct to combine all these forms into a single species.' However, it is also fairly clear that the *I. spuria* complex should be divided in some way since the cytological information available suggests that genetically different plants are involved in the group.

As garden plants, the *I. spuria* variants and hybrids are superb. Much work has been carried out, especially in the United States, to raise a wide range of cultivars of varying habit and flower colour but in Britain they are on the whole perhaps not quite so easily grown and free-flowering. Their flowering period is from May to July.

The following classification of subspecies and varieties should not be taken as a definitive account, but merely as a guide to clarify at least some of the names.

I. spuria subsp. **carthaliniae** (Fomin) Mathew (*I. carthaliniae* Fomin in *Monit. Jard. Bot. Tiflis* **14**: 44 (1909)). This is treated by some authorities as a synonym of subsp. *musulmanica* but I have kept it separate in view of the fact that Dr Rodionenko appears to regard it as distinct. It is a tall plant with large sky blue or white flowers overtopping the leaves, occurring in wet habitats in the southern Caucasus to the west of Tiflis in Georgia. The Russian botanists Soboleva and Rodionenko regard this as one of the finer Spurias, as tall as but of a more delicate appearance than subsp. *musulmanica* and with flowers of a blue rather than violet shade.

I. spuria subsp. **demetrii** (Akhv. & Mirz.) Mathew [*I. demetrii* Akhverdov & Mirzojeva in *Trans. Bot. Inst. Acad. Sci. Armen. S.S.R.* **7**: 27 (1950)]. (Syn. *I. prilipkoana* Kem.-Nat.). This is apparently similar to subsp. *notha* in having dark green leaves and stems 70–90cm in height. The two to five large flowers are violet-blue with darker veins, and white forms are also known. As in subsp. *notha*, the haft of the falls is longer than the blade. It has a more southerly distribution than subsp. *notha*, in Transcaucasia where it inhabits dryish hillslopes, unlike subsp. *musulmanica* which prefers wet places with a high water table.

I. spuria subsp. **halophila** (Pall.) Mathew & Wendelbo (Syn. *I. halophila, I. gueldenstadtiana*). This is one of the more easily recognized variants of *I. spuria* since it has flowers which are primarily yellow in colour. It grows to 40–90cm in height with leaves 0·7–1·2cm wide. The flowers, 6–7cm in diameter, vary from dingy pale yellow to rich golden yellow, usually with darker veining, and they have falls about 4–6cm long with the haft longer than the blade.

Subsp. *halophila* occurs in southern Roumania, the Ukraine, Moldavia, western Siberia and the northern Caucasus and inhabits low-lying wet meadows, often along river banks and frequently in saline soils. Dr Rodionenko makes the observation that it is the most northerly-occurring Spuria iris in Russia and is therefore extremely hardy and useful for breeding purposes.

I. spuria subsp. *maritima* (Lam.) Fournier. (Syn. *I. maritima*). This is the most western variant of *I. spuria*, growing about 30–50cm in height with leaves 0·6–1cm wide. The stem leaves, the upper ones at least, are longer than the stem internodes and conceal them. This helps to distinguish it from subsp. *spuria*, and in addition the bracts are wholly green. It is rather small-flowered with the falls only 3–4·5cm long and the colour is creamy with purple veining, merging into dark blue-purple on the blade. The haft of the falls is longer than the blade and has a greenish-yellow median stripe.

It is a native of south-western Europe, in France and central Spain and possibly also Corsica, growing in damp places.

Forma *reichenbachii* (Klatt) Dykes from North Africa is probably best regarded as synonymous with subsp. *maritima*, or at any rate as a variant of it. It is of robust growth and the stem internodes are not hidden by the leaves to quite the same extent.

I. spuria subsp. *musulmanica* (Fomin) Takht. (Syn. *I. musulmanica, I. violacea* Klatt, *I. daenensis, I. klattii*). In this subspecies the length of the stem is very variable, from 40–90cm, and the stems are thicker and straighter with wider leaves than in subsp. *notha*. The flowers are a bright pale violet to deep lavender-violet, veined darker, with a median yellow signal stripe and sometimes suffused yellow towards the base of the falls, which are 5·5–8cm long. White forms frequently occur in wild populations. The bracts are broadly lanceolate and somewhat inflated. The haft of the falls is about equal in length to, or sometimes a little longer than the elliptical blade.

Subsp. *musulmanica* is widespread in eastern Turkey, northern and western Iran and in Armenia and Azerbaijan. It grows in damp salt marshes and stream-sides and I have seen it occurring in great abundance in Turkey and Iran in such situations.

I. spuria subsp. *notha* (M. Bieb.) Aschers. & Graebn. (Syn. *I. notha*). This has narrow dark green leaves 6mm wide and violet-blue or bright blue flowers carried on slender stems 70–90cm in height which are bent at the nodes. The bracts are not at all inflated, being linear-lanceolate and long-pointed at the apex. The haft of the falls is equal to or longer than the rounded-ovate blade.

It is a native of somewhat drier places than subsp. *spuria*, on well-drained slopes in the steppe foothills of the northern Caucasus. It is considered by Rodionenko to be one of the most elegant and decorative of the Spurias, flowering rather later than most, in July.

I. spuria subsp. *sogdiana* (Bunge) Mathew [*I. sogdiana* Bunge in *Mém. Sav. Etr. Pétersb.* **7**: 507 (1847)]. In the areas east of the Caspian this Spuria predominates. It is shorter than most variants, with stems usually only 10–35cm in height, and has small, 4–6cm-diameter, pale-violet flowers with the falls only 3·5–4cm long. The haft is narrow and about twice as long as the blade. It occurs in the extreme north-east of Iran, Afghanistan, Pakistan, Kashmir and Russian Central Asia, in the Kopet Dag and Tien Shan mountain ranges, and east to Mongolia. The usual habitat is in wet meadows or along irrigation ditches.

Subsp. *sogdiana* is an unimpressive plant, one of the least attractive of its group, but its compact habit might make it a useful parent in breeding shorter cultivars.

I. spuria subsp. *spuria*. This is usually about 50–80cm in height and has 8–12mm-wide leaves about equal in height to the flowers. The stem leaves, at least the upper ones, are shorter than the internodes of the stem so that parts of the stem are visible. The bracts are semi-transparent at the apex and they enclose two to four flowers, each about 6–8cm in diameter. These are lilac with violet veins and a yellow median signal stripe on the falls which are about 4·5–6cm in length. The haft of the falls is considerably longer than the elliptical or orbicular blade.

Subsp. *spuria* is a central European plant, occurring in wet, usually saline soils in southern Sweden, Denmark, Germany, Austria, Hungary and Czecho-slovakia. It is also recorded in England in Dorset and on the Lincolnshire coast.

Var. *danica* Dykes is morphologically very similar.

Var. *subbarbata* (Joó) Dykes was so named because of the presence of unicellular hairs on the falls. This is not comparable, however, to the beard on a pogon iris and there is no need to doubt the classification of the genus *Iris*! Var. *subbarbata* appears to be eastern European, mainly occurring in Roumania, Hungary, southern Germany and eastern Austria.

I. stenogyna Delarbe. This is a synonym of *I. spuria*, probably subsp. *halophila*.

I. subbarbata Joó. A synonym of *I. spuria* subsp. *spuria* var. *subbarbata*.

I. 'Turkey Yellow'. This is a yellow-flowered Spuria from Turkey growing about 50–100cm in height with the 1–1·8cm wide leaves shorter than or equalling the flowers. The flowers are wholly deep yellow, about 9–11cm in diameter, the falls with an orbicular, broadly elliptical or sometimes ovate blade about 3–4cm long and equalling the haft in length. In these respects it differs from *I. crocea* which has much larger flowers with a more oblong blade longer than the haft. Furthermore the small style branch lobes of the Turkish plant are very strongly recurved, much more so than in *I. crocea* or *I. orientalis*.

This has been collected in several places in central and southern Turkey, for example near Muğla, in the south-west, Antakya in the south near the Syrian border and near Ankara. It grows in marshy meadows or streamsides from as low as 40 metres altitude to probably around 1000 metres and it flowers in April or May. Dr Lee Lenz of the Rancho Santa Ana Botanic Garden introduced material of this plant to America in 1948 and it is being successfully used in the breeding of new Spuria cultivars. He has suggested the name 'Turkey Yellow' in the apparent absence of a latin name, until such a time as it can be studied more fully and formally described, that is if it is considered to be a distinct species.

I. urumovii Vel. This is obviously very similar to *I. sintenisii* and is, in the *Flora of Bulgaria*, regarded by Stojanov and Stefanov as a synonym of this. It is a very narrow-leaved plant (leaves only 1–2mm wide) but this is within the normal range of variation of *I. sintenisii*.

I. violacea Klatt. A synonym of *I. spuria* subsp. *musulmanica*.

2B(l)   Subgenus Limniris, Section Limniris, Series Foetidissimae

This series contains only *I. foetidissima* which may be instantly distinguished from all the other beardless irises by its bright red seeds which remain attached to the capsules for a long time after they have split wide open. It is not a showy plant in flower, but is useful in the garden since its tufts of leaves are evergreen and the scarlet seeds are a welcome sight in the depths of winter. It is easily propagated by seed, or the clump can be divided in spring or autumn. It has no special requirements and will grow in almost any soil or in any position from full sun to quite deep shade. Unfortunately it is very susceptible to virus diseases causing yellowish streaking in the leaves and malformed flowers. However, new plants raised from seed are virus-free.

I. foetidissima Linn. The Roast Beef Plant, Gladwyn or Gladdon iris. This well-known plant carries these curious local names in Britain, the first referring to the smell of the bruised leaves, although I must admit that I cannot find any great resemblance to the aroma of roast beef. The other names are probably derived from the latin 'gladius', a sword, in reference to the leaf shape. It is a variable plant usually about 30–90cm in height with evergreen tufts of tough dark green basal leaves about 1–2·5cm wide, produced from compact tough rhizomes. The rather flattened stem has two or three branches, each bearing one to three flowers opening in succession. These are 5–7cm in diameter and are usually a dull purplish-grey tinged with dull yellow. The plant which I grow is fairly typical of the species and this has a lilac-blue lamina to the falls, strongly reticulate-veined purple and fading to white in the centre and shading to brown on the haft. The rather smaller standards and styles are brownish, the former faintly flushed lilac in the centre of the blade.

The flowers have a slightly disinfectant scent very reminiscent of that curious Californian Trillium-like plant, *Scoliopus bigelovii*, which is pollinated by flies.

There are also yellow-flowered variants, for example var. *citrina* Syme, which has been collected in the Purbeck area of Dorset, and var. *lutescens* Maire, a clear yellow form from Algeria. Yellow forms are also known from the Isle of Wight and in Sicily. There is in some of these yellows a shading of brown on the hafts of the falls and standards. Although the overall flower shape of *I. foetidissima* has nothing particularly distinctive about it, the haft of the falls is notable in being winged and sometimes nearly as wide as the blade, from which it is separated by a constriction. The perianth tube is only about 1cm in length and is rather broad and cup-like. As mentioned above, the capsules split widely open to reveal the scarlet seeds which hang on for several months, so that they are useful for picking for winter decorations. A white-seeded form was known at one time, found in 1921 in a hedgerow among ordinary red-seeded forms by Mary Ellin Shedden of Bridgewater, Somerset. I have also heard of a yellow-seeded form but have seen neither of these variants.

I. foetidissima is a widespread plant in western Europe from Portugal to Italy, Sicily, Corsica and Sardinia, and northwards as far as Scotland and Ireland. It is also in Morocco, Algiers, the Azores and the Canaries. The North African plants obviously belong to the same species but can be very vigorous, up to 1 metre in height with flowers up to 9cm in diameter. In Britain it is usually a plant of chalk and limestone areas in woods and scrub or on cliffs near the sea. The normal flowering period is in June or July.

In addition to the plain dark-green leaved form there is a variegated one which is a useful foliage plant for a dull corner of the garden.

2B(m) Subgenus Limniris, Section Limniris Series Tenuifoliae

An Asiatic group of iris, the Tenuifoliae have small near-vertical rhizomes producing a tufted habit, not widely spreading. The narrow leaves are tough and rather like those of the Spuria irises. A distinctive feature is the way in which the bases and fibres of the old leaves persist as a neck at the apex of the rhizomes. Their flowering and fruiting characters are mainly used to distinguish them from other series in the section: the capsule is cylindrical or ellipsoid, sometimes with six strong ribs, one on each face in addition to those at the three corners, and it contains wrinkled cubical or rather pear-shaped seeds which have no fleshy appendage. They are distinct from the D-shaped seeds of the Spuria irises in which there is a smooth loose shiny coat which separates easily from the body of the seed. The flowers have a stigma which is bilobed (in several other groups it is like a triangular tongue) and the perianth tube is usually quite long, usually 4–14cm in length. In this last feature the Series Tenuifoliae is therefore similar to the Series Unguiculares but the latter has a quite different horizontally creeping rhizome and is ecologically and geographically quite separate.

The Tenuifoliae are mostly plants of open steppe and semi-desert country in central Asia. I have included in the Series, with *I. tenuifolia*, several other species such as *I. ventricosa* which G. I. Rodionenko separates into yet another group, the Ventricosae.

The Tenuifoliae irises

I. bungei Maxim. A distinctive species, with the stems 15–30cm in height and producing rather inflated bracts 8–10cm long; these have delicate, more or less parallel veining and become straw-like in texture in the fruiting stage. The habit of growth is similar to that mentioned above, rather tufted with the old leaf bases persisting as a coarse neck. The leaves, some of which are carried on the flower stem, are about 2–5mm wide. Although I know this only as herbarium specimens it is clear that the flowers are about 5–6cm in diameter and have fairly narrow falls and standards, and several of them emerge from each set of spathes. The colour is probably similar to that of *I. tenuifolia* (see below). In fruit it has large cylindrical capsules with a beak 5cm long and has a strong rib on each of the three faces of the locules as well as those of the three corners.

I. *bungei* occurs wild in northern China and eastern Mongolia where it flowers in June. Some years ago it was said to be abundant on plains and mountainsides there. It does not, unfortunately, seem to be in cultivation.

I. cathayensis Migo. This probably belongs in the Series Tenuifoliae although I have seen only a drawing. This shows tough, almost erect rhizomes with persistent leaf bases and narrow rigidly erect leaves overtopping the flowers which are short-stemmed but long-tubed and with very narrow segments. The flowers have falls with an ovate blade narrowing abruptly to a very narrow haft, narrowly oblanceolate erect standards, and narrow pointed style branches. The falls appear to be spotted on a pale ground but I do not know the colours. The bracts are long and acute, not inflated at all which would distinguish it from *I. bungei* and *I. ventricosa*, but one would like to see some specimens to identify the differences between it and *I. tenuifolia*. It is a native of China.

I. kobayashii Kitagawa. This is classified here with the Tenuifoliae on the basis of information available in literature. It is tufted and probably about 10–15cm in height at flowering time, from a short near-vertical tough rhizome which has strong fleshy roots and persistent leaf bases. The erect rigid leaves are longer than the flowers and about 2–3mm wide. There are two flowers per stem, which is very short, but they have slender perianth tubes about 5cm in length, funnel-shaped at the apex. They are probably around 5cm in diameter and are said to be yellowish, densely purple-spotted, and have an elliptical, horizontal lamina to the falls which narrows abruptly to the haft. The standards are erect and narrowly oblanceolate and the stigma flap is described as having two obtuse lobes which would be acceptable for Series Tenuifoliae. In fruit there is a six-ribbed capsule with a beak at the apex, and the seeds are wrinkled, giving further strength to the idea that this belongs with *I. tenuifolia*. *I. kobayashii* grows in northern China (Manchuria) where it flowers in April and May.

I. loczyi Kanitz (Syn. *I. tianschanica*). A dwarf plant, sometimes regarded as a variant of *I. tenuifolia* although it does seem to be slightly different and fits into a separate ecological niche; it is densely tufted, forming clumps up to 40cm in diameter, and is usually 15–30cm in total height. The fibrous dead leaf bases are very persistent and form very hard tussocks. The leaves are about 10–30cm long and 2–4mm wide, generally wider than those of *I. tenuifolia*. For the size of the plant the flowers are relatively large, about 4–6cm in diameter, and they have a very long tube usually 10–14cm long. Although there is only a very short stem present, this tube pushes the flower well up from the ground. The colour of the standards and styles is a very pale blue or bluish-purple and the falls are veined purple-blue on a pale yellowish-cream or white ground with a more yellow tinge in the centre. The capsule is nearly stemless, about 2·5cm long and rather dumpy with a slender beak. It is a plant of the mountains, occurring on open sunny rocky or grassy slopes usually at 2200–2600 metres and flowering in April or May. The area of distribution is from north-east Iran, through Afghanistan into Russian Central Asia in the Tien Shan and Pamir Alai mountains southwards into Baluchistan. I have also seen specimens, which appear to be this species, from Tibet at the very high altitude of 3500–3800 metres.

I. regelii Maxim. The specimen I have seen of this looks like a miniature *I. tenuifolia* only 6–7cm in height to the top of the flowers, which are only about 3cm in diameter. It was described from Russia, in Turkmenistan.

I. songarica Schrenk. A common semi-desert plant distributed over an enormous area of central and eastern Asia. It is up to 80cm tall and is clump-forming, appearing almost like tussock grass when out of flower and making tufts 60cm or more across. The tough leaves are often taller than the flower stems and only 1–3mm wide, some of them shorter and sheathing the stems. The flowers are about 5–7cm in diameter, with up to six from each set of bracts and one to four branches on each stem, so a whole clump can be very floriferous, although the soft greyish lavender-blue colouring is not very striking. The whole flower is lightly spotted and veined darker, giving the grey appearance. In fruit, it has a rather cylindrical capsule 4·5–6cm long with a beak at the apex but without the strong ribs of *I. bungei* and *I. ventricosa*.

I. songarica is the tallest member of its group, so robust that it is immediately distinguishable from all the others. The bracts are not inflated as in *I. bungei* and *I. ventricosa*. It is a native of sandy plains and dry rocky situations at altitudes of 900–2500 metres in Iran, Pakistan (Baluchistan province), Afghanistan, Soviet Central Asia, Mongolia and northern China.

I have never succeeded in growing *I. songarica* to more than the seedling stage, the seeds kindly supplied by John Ingham. Plants sent back by him died within a short space of time and I think it is nearly impossible to establish the species in cultivation by this means. The wiry rhizomes have little in the way of reserves to overcome the shock of transplantation. Obviously seeds give the best chance of success, but the seedlings seems to rot off in winter at the slightest suggestion of humidity in the air.

I. tenuifolia Pallas. A widespread species and the best known of its series. It is a densely tufted plant with a mass of brown fibrous leaf remains around the base of the stem. The height is usually about 10–30cm but the narrow, rigidly erect leaves can reach 40cm and are only 1–3mm wide. There are one or two shortly stemmed scented flowers about 4–6cm in diameter carried within long bracts which are usually about 7–10cm in length. The perianth tube is 5–8cm long with the ovary well down in amongst the leaf bases where it remains right through to the ripe capsule stage. Colour variation is not very great, usually ranging from a bluish-violet to paler lilac with the falls tending to be heavily veined on a creamy ground, rather than wholly violet coloured. There is a creamy or yellowish central band on the falls but this is not raised into a ridge. Both falls and standards are narrow and often rather pointed, only 1–1·5cm wide across the blades, and the style branch lobes are very narrow and sharply pointed. The capsules are short and dumpy, at most 4·5cm long including a short beak.

I. tenuifolia is a plant of open sandy or stony steppe country, or sandy river banks, from south-eastern Russia (Kazakhstan) eastwards through Russian Central Asia to Mongolia and western China. It flowers in April or May in the wild and occurs at altitudes of usually about 1000–2000 metres, although in the east of its range it exceeds this.

In cultivation in Surrey it grows well in full sun on top of a bank of sandy soil but is not very free-flowering and the soft lavender flowers tend to be lost among the tufts of greyish-green leaves.

Although *I. tenuifolia* and *I. loczyi* are obviously very closely related, the former tends to be a plant of semi-desert steppe and the latter, rocky mountain slopes at higher altitudes. In general the leaves of *I. loczyi* are shorter and broader and the perianth tube extremely long (10–14cm).

I. tianschanica (Maxim.) Vved. ex Woron. A synonym of *I. loczyi.*

I. ventricosa Pallas. This is rather similar to *I. bungei* but has even more inflated bracts laced with a delicate network of veins, not only parallel ones. It grows about 8–20cm in height, the flowers overtopped by the tough 3–7mm wide leaves. The wiry rhizomes produce rather thick fleshy roots, judging from the dried specimens I have seen. The falls and standards are rather narrow and appear to be widespreading, giving the flower a diameter of about 5–7cm and a rather 'spidery' appearance. It is impossible to judge the colour from herbarium material but one record notes 'pinkish-blue', so it is probable that the flowers are lavender colour similar to those of *I. songarica.* Like *I. bungei,* the species has an elongated capsule with a long narrow beak at the apex and six prominent ribs.

It inhabits sandy plains and river banks in Mongolia, north-western China and south-eastern Russia, in the Ussuri region. I am unfortunately not familiar with living plants. The species would be worth cultivating for the curiosity of its large inflated bracts alone.

2B(n) Subgenus Limniris, Section Limniris, 'Series Ensatae'

The Series Ensatae has an unfortunate name, since it has long been known that *Iris ensata* of Thunberg is the correct name for the Japanese Water iris widely grown as *I. kaempferi* (Series Laevigatae). The name was misapplied, to an entirely different widespread non-aquatic iris from central and eastern Asia which should be called by the earliest available name of *I. lactea* Pallas (1776). It is a very variable plant, and botanists have described different variants over and over again as species so that there are now several synonyms. On the evidence of a lot of herbarium material there seems to be no reason to divide up *I. lactea*, for the variations such as habit, flower colour or flower shape do not seem to tie up sensibly with other factors such as geographical distribution or habitat.

The main features which define this series concern the long narrow beaked ovary which is separated from the perianth segments by a tube only 2–3mm long, so that the falls and standards appear almost free from each other. The ovary is clearly six-grooved and gives rise to the beaked cylindrical or oblong capsule with six prominent ribs and containing shiny globose seeds. The plants are clump forming, have narrow leathery leaves and are known to be salt resistant, often growing at the margins of irrigation ditches and in marshes. In some places in Central Asia *I. lactea* is considered as a useful fodder plant since it will grow in places where many other plants will not. Also the leaf fibres are so tough that they have been utilised to make a string or rough cloth. The resistance of this plant to its harsh environment is partly due to the tough deep roots which enable it to survive where other more shallow rooting plants cannot.

Cultivation

I. lactea presents no problems in gardens providing it is given an open position and reasonably well-drained soil. It survives long periods of drought and flowers quite freely, although it can hardly be regarded as a showy garden plant.

I. lactea Pallas (Syn. *I. biglumis* Vahl; *I. fragrans* Lindl.; *I. iliensis* Pol.; *I. moorcroftiana* Wall. ex Don; *I. oxypetala* Bunge; *I. pallasii* Fisch.; *I. triflora* Balbis; *I. ensata* Dykes [and many other authors but not of Thunberg] and *I. lactea* var. *chinensis* (Fisch.) Koidz.) This is a very variable clump-forming species 6–40cm in height with tough wiry-rooted rhizomes covered with reddish or purplish-brown fibrous remains of old leaf bases. The 3–5mm-wide erect grey-green leaves equal or overtop the flowers and are very tough with strong ribs. Each stem carries two or three fragrant flowers about 4–6cm in diameter, carried on pedicels 2–4cm long, elongating to 5–7cm at fruiting time. The flower shape is fairly distinctive because the falls do not spread widely but form a narrow V-shape. The colour is primarily blue, bluish-violet or purple, the falls usually with a paler, whitish or yellowish dark-veined haft. White

forms occur in mixed populations with the blue according to some field
observations. The perianth tube is very short, only 2–3mm long, but there is a
slender beak at the apex which might be mistaken for a tube but for the fact
that it is solid, not hollow, and remains attached to the ovary after the flower
has faded. The perianth segments are all rather long and narrow thus giving
the impression of a flower of little substance. Both falls and standards are
oblanceolate, the former with a blade up to 2cm wide and the latter more
narrow and shorter. It is the comparative size and shape of the segments which
has led some authorities to 'split' *I. lactea* into two species (*I. lactea* and *I.
oxypetala*). Vvedensky in *Conspectus Florae Asiae Mediae* (1971) gives *I. lactea* as
violet or white with the falls blunt at the apex and much broader than the
standards, while *I. oxypetala* is said to have blue, more or less equal, pointed
falls and standards. Added to these characters are the shapes of the capsules,
broadly oval (ovoid) and shortly-beaked in *I. lactea* and narrowly-cylindrical

SUBGENUS LIMNIRIS, SERIES ENSATAE: *I. lactea*

with a long beak in *I. oxypetala*. This may be correct, although from the fairly wide range of herbarium material available I have not been convinced, taking into account specimens from the whole range of distribution. The capsules vary from rather short, fattish oblong ones to long-cylindrical ones, from 4·5–8cm long and 0·7–1cm wide. The size, shape of segments, habit of the plants and flower colour do not seem to be associated with the distribution or a particular habitat, although the specimens from western China tend to be short and have large flowers with the leaves only just overtopping them.

I. lactea occurs in Russia from Kazakhstan through Soviet Central Asia to the far-eastern Ussuri region. It is in Korea and is widespread in central, north and western China and Mongolia, and in the Himalayan region occurs from the north-eastern border of Afghanistan with Pakistan through Kashmir and the Punjab to Tibet. The habitat varies from open dried up river beds to banks between fields, marshy places and sandy lake-margins, at altitudes of 600–3700 metres. It flowers from May to August depending upon altitude but in Britain usually blooms in May or June.

2B(o) Subgenus Limniris, Section Limniris, Series Syriacae

A distinctive Series, the Syriacae are separated from all others by having a more or less vertical tough rhizome with rather swollen leaf bases, giving a bulb-like appearance, and a neck of very spiny bristles in a cluster at its apex. These spines are stiff and needle-like and are capable of inflicting painful damage to fingers at planting time!

The number of species is open to some doubt. I would hesitate to lump them all together as one variable species since certain features seem to coincide with distribution and/or habitat. Here, therefore, three species are definitely accepted, another poorly known one of uncertain origin is included.

Cultivation

To date I have been unsuccessful in flowering *I. masia*, the ony one I have had the opportunity to try. It has grown well in a pot, in a bulb frame and in an open sunny border against a fence but although strong healthy growths appear each year the clump shows no inclination to flower. This seems to be the experience of other growers also. The next experiment will be to try heavy feeding, especially with high potash fertilizer which often encourages bud initiation in bulbous plants.

The Syriacae irises

I. aschersonii Foster. This was described from material received from W. Siehe of Mersin, said to have originated near Adana in southern Turkey. I have seen no material collected in that area although it is not far to the west of the most westerly records for *I. masia*. It was said to have greenish-yellow flowers, the falls of which were lined with thin veins and linear blackish dots, sometimes also spotted in the centre of the blade. The leaves were said to be much narrower than those of *I. grant-duffii*, which of course they are in *I. masia* and *I. melanosticta* also. It is possible that *I. aschersonii* represents a colour variant of *I. masia* with more green in the flower, and it may be that these two and *I. melanosticta* are colour forms of one narrow-leaved species from inland dryish areas, leaving *I. grant-duffii* as a distinct but close relative. Until further field studies can be made I feel that the question must remain unresolved.

I. grant-duffi Baker. This grows about 25–30cm in height from a rhizome as described above, and with stiff narrow grey-green leaves up to 35cm long and 5–10mm wide, ribbed with many prominent veins. The solitary flowers are faintly primrose-scented and are about 6cm in diameter. They are basically suphur- or greenish-yellow, sometimes with a few black dashes on the blade of the falls and with an orange signal patch on the centre. The haft of the falls is often veined purplish on a whitish or pale yellow ground and the standards are either the same yellow as the falls or paler. It has a much more substantial flower than that of either *I. melanosticta* or *I. masia* in which the falls and standards are both shorter and narrower. In *I. grant-duffii* the falls are 6–7cm long and their haft is 1·1–1·2cm wide; the standards are about 7cm long. In the other two species the falls are 5–6cm long and 5–7mm wide on the haft, and the standards are about 5–6cm long. *I. grant-duffii* gives the whole impression of being a stockier plant with shorter, broader leaves. It is a plant of the lowland Mediterranean region of Israel, Syria and probably the Lebanon, growing in coastal swamps and damp fields on heavy loam, and flowering in February and March. It is known from the Haifa Bay region, the banks of the river Kishon, the plain of Esdraelon and to the south of Latakia in Syria.

I. masia Stapf ex Foster. (Syn. *I. caeruleo-violacea*). The rootstock is the same as for *I. grant-duffii*, and the general habit of growth, but it varies from 25–70cm in height and has six to eight rigid leaves up to 60cm long and and only 3–5mm wide. The flowers are violet-blue with distinct veining, especially on the falls which have a whitish ground colour in the centre and on the haft. All the flower parts are narrower and slightly smaller than those of *I. grant-duffii* (see measurements given above), producing a less substantial-looking flower. It flowers in April and May in open fields or in dryish steppe on basalt formations at about 750–1100 metres in south-eastern Turkey, north and central inland Syria and adjacent Iraq. Thus it is a plant with a considerably different habitat from that of *I. grant-duffii*. *I. masia* is named after the extinct volcano Karacadağ (the 'Black Mountain') in southern Turkey, whose ancient name was Mons Masius.

I. melanosticta Bornm. This resembles *I. masia* quite closely in general habit of growth and dimensions of the flower parts, but it has yellow flowers with a few large linear black marks on the blade of the falls. The yellow ground is a much purer yellow than that of *I. grant-duffii* which is of a greenish shade, and the falls and standards of the latter are broader and longer. It is an inland plant of dry basalt country in southern Syria and possibly northern Jordan, flowering in March and April.

2B(p)   Subgenus Limniris, Section Limniris, Series Unguiculares

A very well known group of irises, frequently cultivated and much-loved for their mid-winter flowers; they have tough rhizomes, branching very freely to form large patches which eventually die out in the centre leaving a circle of growth. Several wiry evergreen leaves are produced from each growing point. Although the flowers (and later, the capsules) are stemless, they have an extremely long perianth tube which is long enough to enable them to be used as cut flowers. The standards and falls are well-developed, as in most irises, but the falls have an orbicular or oblong-elliptic blade and a long narrow tapering haft. The style branches have two features which are rather distinctive. They are joined at the base into a very distinct slender tube, this having suggested the name *Siphonostylis* to W. Schulze who created a separate name for *Iris unguicularis* and its relatives. They also have yellow glands on their margins, which Dykes likened to a sprinkling of gold dust.

Cultivation

For *I. unguicularis* and its variations, a hot sunny spot in the garden is the best that one can provide, for the plants need considerable warmth in summer if they are to flower freely. Good drainage is necessary and they are more satisfactory on alkaline than on acid soils. Although they should not be given a rich diet, they do respond well to dressings of potash or bone meal in the autumn and spring. It is sometimes stated that the plants flower better if their foliage is shortened by clipping, thus allowing more light on to the rhizomes. I have not found this to be helpful but I do make sure that dead leaves are removed from the clumps, with the same purpose in mind.

Propagation is by division of the clumps in autumn or spring. I prefer the latter since it takes place just after flowering and therefore does not spoil that season's flowers, but one must be careful to avoid any drying-out of the soil until the transplants are well established.

I. cretensis (here treated as a variant of *I. unguicularis*) is much more difficult to flower in British gardens and is undoubtedly better in a bulb frame where its rhizomes get a thorough summer baking. *I. lazica* on the other hand prefers cool places in semi-shade, although I have also seen it doing well in full sun.

The Unguiculares irises

I. unguicularis Poir. (Syn. *I. stylosa*). This really needs little introduction since almost every gardener, even if not an iris enthusiast, knows it for its delightful primrose-scented lavender flowers in winter, from November to February. In the most 'typical' form the tough linear leaves are about 1cm wide and

45–60cm in length. The large flowers, 5–7cm in diameter, with falls 7–8cm in length, have an exceedingly long perianth tube, 6–20cm, and are almost stemless so that the developing ovary is buried right down among the leaf bases and does not become visible until it is a ripe, but still stemless, capsule. The most usual form has flowers of a uniform mid-lavender with a yellow signal band in the centre of the falls and with dark veining on the haft. The form we know in gardens comes from Algeria and Tunisia, while in Greece, Crete, western and southern Turkey, Syria, Rhodes, Lebanon and possibly Israel (see reference below to Lake Tiberias), there are various other forms, one of the most notable of which is 'Cretensis' (see below). In the wild, *I. unguicularis* is usually a low-altitude plant growing in scrub and rocky hills and flowering normally in early spring.

Apart from the North African form, which we may regard as 'typical' of the species, there are many others but it is difficult to give them any formal names, such is the great variability. In south-west Turkey one often sees a very vigorous form with leaves up to 60cm long but only 2mm wide. On the other hand one can also find in the same area dwarf plants with leaves only 7cm long, and sometimes forms with leaves as much as 5mm wide. I have also seen narrow-leaved forms from North Africa, so it appears that one cannot satisfactorily 'break up' *I. unguicularis* into more than one taxon. A curious collection by G. P. Baker was made on the 'upper slopes above Lake Tiberias'. This has very robust leaves about 1·8cm wide and large flowers with falls 9cm in length. The one point which these variants seem to have in common is the large, predominantly lavender flowers. There is some variation in depth of colour and amount of veining and markings on the falls, but not such striking colour differences in the parts of the flower as in the small-flowered 'Cretensis'.

There are many named selections of *I. unguicularis* and no attempt is made here to account for them all since doubtless some of the older ones are either now lost to cultivation, or at least of uncertain identity. There are in addition some cultivars in the United States which I have not seen.

'Alba'. There are said to be several whites, some better than others. The form I have is very floriferous but has rather narrow perianth segments. From time to time I have been given 'good white forms' but they have always turned out to be this same one! The colour is a pure creamy-white except for the contrasting greenish yellow median line on the falls.

'Mary Barnard'. This is said to have come from Algiers. It is a good violet blue form, currently obtainable in Britain.

'Speciosa'. A short- and narrow-leaved variety with deep velvety violet flowers with a strong scent of honey. The falls have an orange stripe in the centre.

'Angustifolia'. Although the flowers are rather small they are interestingly coloured, the falls being white in the centre with bluish-lilac margins. The standards also are white near the base.

'Walter Butt'. A robust large-flowered form, very fragrant, in a pale silvery-lavender.

'Variegata'. A curious form in which the flowers are mottled and streaked purple on a lavender ground colour. It is, to my mind, not very pleasant and suggests a virus infection.

'Starkers Pink'. A very dwarf plant with short narrow leaves and pinkish-lavender flowers. It is very shy flowering and rather weak, my plant having died without flowering. I suspect it to be a form of 'Cretensis' and if I do obtain it again will plant it in a bulb frame.

'Oxford Dwarf'. This looks exactly like plants I have seen growing wild in the Peloponnese and I imagine it to be a form of 'Cretensis' (see below). The falls are tipped with lavender but are predominantly white, veined purple with an orange line in the centre. The segments are narrow and therefore widely separated. The material I saw was grown at Kew, and came from Oxford Botanic Garden.

'Ellis's Variety'. A narrow leaved form with rich violet-blue flowers.

'Marginata'. This has lilac flowers with an edging of white to the falls and standards.

In addition to these named forms there are many other variants, including a really good pink which I have seen only once.

I. unguicularis 'Cretensis' (Syn. *I. cretica*, *I. cretensis*). This dwarf iris can probably be recognized as a distinct entity, although in size alone it cannot be differentiated, for I have seen specimens of *I. unguicularis* from Turkey and North Africa which are just as small. It appers that 'Cretensis' is restricted to Crete and the Peloponnese and may be recognized by its short, narrow leaves, usually only 1–3mm wide, and by the smallish strikingly coloured flowers with narrow segments. The falls have a violet or deep lavender apex but the rest of the blade, and the haft, is white with violet veins. In the centre is an orange median stripe. The standards are lavender or violet with a brownish, violet-spotted zone on the haft, while the small style branches are a pale bronzy-purple. The segments are all very narrow, with the oblong-elliptical blade of the falls only 1–1·5cm wide and the standards less than 1cm wide. The hafts too are only a few millimetres wide, so the flower parts are all widely separated, leaving the column formed by the joined style branches clearly visible in the centre. The falls and standards do not exceed about 5·5cm in length whereas in subsp. *unguicularis* they are usually about 7–8cm. 'Cretensis' grows in open rocky places and dwarf scrub, usually on limestone, in Crete and southern Greece and flowers in February, March or April. It is not easy to get it to flower freely in Britain and it does best in a bulb frame. However, a clump in one chalky garden on the South Downs flowers extremely well every year with no protection.

In May 1980 I had the opportunity to see some fresh seeds collected in Crete by my colleague Phillip Cribb. These were nearly spherical, about 6mm in diameter, and covered with glistening sessile glands. Some ants (on the floor of my room in the herbarium at Kew!) were strongly attracted to them and it is fairly certain that this is the natural method of seed dispersal in this group of irises, whose capsules nestle among the leaf bases. Without some dispersal mechanism such as this the seeds would be released in a very inefficient way, into the crown of the parent plant.

I. lazica Alboff. This is a very different plant in appearance and habitat from *I. unguicularis* and there is no doubt that it should be treated as a distinct species. It grows about 15–25cm in height and has broad bright green leaves up to 1·5cm wide in distinct spreading fans, rather than compact erect tufts as in *I. unguicularis*. The flowers are predominantly lavender-blue but the lower half of the falls is white, spotted and veined with darker lavender. These spots continue out on to the blade of the falls and there is a pale yellow median stripe in the centre. The standards are lavender-blue and unspotted. From the plants I have collected and dried specimens seen in herbaria, it would appear that *I. lazica* has a generally shorter tube—about 7–10cm—than *I. unguicularis*, although there is some overlap in measurements. It is sometimes stated that it has a short stem, but I have never seen a stemmed form. *I. lazica* occurs in the damp climate of the Black Sea coast regions of south-western Caucasia and north-eastern Turkey as far west as Giresun. It is March-flowering and is a low altitude plant, from sea level to about 400 metres, growing on shady banks in mixed scrub or hazel nut plantations, often with *Rhododendron ponticum*. It is thus in damp conditions for most of the year, an entirely different habitat to the sunbaked hillsides where one finds *I. unguicularis* and its variant 'Cretensis'.

In Britain, *I. lazica* is an easy plant to grow and I find it best in slight shade where it flowers freely, rather later than *I. unguicularis* and its forms.

3   Subgenus Nepalensis

This contains one very distinct, fairly recently described, species and two others which can be regarded as distinct enough to separate into species, although I am not familiar with one of them, *I. collettii*, in the living state. The large amount of herbarium material of this and *I. decora* is easily separated into two batches using one or two obvious characters, notably the stemless habit of *I. collettii* and its smaller flowers. This does not seem to be associated with high altitudes or exposure since both species can be found in full sun or amongst other vegetation over a similar altitude range. In fact *I. decora*, the taller of the two, can occur as high as 4300 metres, and *I. collettii*, the dwarf species, as low as 1300 metres and possibly even down to 550 metres (see comments below). Their distributions are different, meeting only in south-western China.

The features of the Nepalensis Subgenus, which is recognized as a separate genus *Junopsis* by some authorities, are associated mainly with the rootstock. This consists of a small growing point to which the plant dies back completely in winter like many herbaceous plants, and attached to this are some storage roots which are either swollen up to the point of attachment or thin to begin with and then swollen towards the ends. At the apex of the growing point is a tuft of fibrous leaf bases, left over from several previous years. The leaf veining is very prominent in the fresh state. The flowers are very short-lived, each lasting less than one day, and they are either uncrested *(I. staintonii)* or have a linear crest on the falls which is not unlike the cockscomb-like protruberance of the evansia iris.

Cultivation

I. decora, the only species in cultivation, presents little difficulty, but does not persist for long in the open ground. A deep pot in an alpine house or cold frame suits it best, where the plants can be nearly dried out in winter when dormant. In the summer it prefers cool growing conditions with plenty of moisture through to flowering and fruiting time in June to August. New plants are easily raised from seeds which are produced fairly readily.

The Nepalensis iris species

I. collettii Hooker. (Syn. *I. duclouxii*). This is similar in many respects to *I. decora*, in its rootstock with bristly fibres and in its leaf features. It is however a dwarf plant at flowering time with stemless or near-stemless flowers. The whole plant to the top of the flowers is often only 5cm in height but the leaves usually elongate after flowering, leaving the capsules, however, on very short stems. Although nearly stemless, the flowers have an extraordinarily long perianth tube, up to 10cm in length, but in very small specimens it is obviously much shorter than this. The diameter of the flower is only 2·5–3cm. They are fragrant and of a pale bright blue with a yellow-orange crest, and are produced either singly or two to each set of spathes. Often there are several inflorescences per plant so that there can be more than one flower out at a time. My impression, from herbarium specimens, is that it can be a very floriferous, and probably beautiful, little plant. It occurs in south-west China in Szechuan and Yunnan provinces and into mountainous northern Burma and is a plant of open grassy sites and in scrub or open pine forest at 1300 to 3400 metres, flowering from April to July.

There is a similar plant from northern Thailand and I think this belongs with *I. collettii* although it has much more robust leaves up to 1·5cm wide. It is otherwise a fairly dwarf stemless-flowered plant but with a very leafy appearance compared to *I. collettii*. It occurs at fairly low altitudes, from 550–1000 metres in open pine or deciduous forests.

I. decora Wallich (Syn. *I. nepalensis*, *I. yunnanensis*). This is best known as *I. nepalensis*, but unfortunately the name was first used for a tall bearded iris and cannot be used again, even though the 'other' *I. nepalensis* is not a good species. *I. decora* has a distinct stem which varies in height from 10–30cm and has the rootstock features mentioned above. The erect strongly ribbed grassy leaves are about 2–5mm wide and overtop the flowers which are produced on branched or unbranched stems. They are slightly scented, about 4–5cm in diameter, pale bluish-lavender to a deep reddish purple and whitish veined with purple on the haft of the falls. The crest is bright yellow-orange up to the apex where it either changes to white or purple. The style branches have broad lobes which are crisped and toothed at the margins. There is a rather distinctive shape to the whole flower since the standards are not erect but bend outwards and slightly downwards to the same degree as the falls. This gives an overall flattish and almost regular appearance, although the falls are larger than the standards and of course are quite distinct because of their yellow crest. The perianth tube is about 3·5–5cm long.

I. decora is a native of the Himalayas over a very wide area from Kashmir in the west through Nepal, Sikkim and Bhutan to the south-west China provinces of Yunnan and Szechuan. To the north it reaches into southern Tibet and to the south into the north Indian mountains of Simla, Garwhal, Khasia and Shillong. It is a plant of open sunny hills and pastures, mountain scrub, rock crevices or clearings in rhododendron forest, flowering in May to July at altitudes of 1000–4300 metres.

I. duclouxii Lév. A synonym of *I. collettii*.

I. leptophylla Lingelsheim. This was described from material collected in Szechuan, China, by Limpricht in 1922. I have seen the original specimen and it appears to be a vigorous *I. decora* with flowers 5–6cm in diameter.

I. nepalensis D. Don. A synonym of *I. decora* Wall.

I. staintonii Hara. This most interesting iris was discovered by J. D. A. Stainton and was described in 1974. Unfortunately living material was not introduced and we must await further collections from central Nepal.

It has an extraordinary rootstock consisting of a very small rhizome, just a mere growing point really, to which are attached some thin fibrous roots and one, or sometimes two, small but fat tubers about 1–1·5cm long. The stem is about 3–8cm tall and has only one well developed linear-ensiform leaf which is 15–30cm in length and about 4–6mm wide. The other leaves are short, blunt and rather bract-like. There is one small flower per stem, about 3cm in diameter with a tube 2–2·5cm long, and it is pale violet with no crest on the falls. This absence of a crest, and the size of the flowers, distinguishes it from *I. decora*, while the short tube and longer flower stems separate it from *I. collettii*.

I. staintonii was collected in alpine meadows at 3800 metres in the central Nepalese Ganesh Himal, flowering in July.

4   Subgenus Xiphium (The 'Spanish irises')

These are the tall bulbous irises of western Europe and North Africa which we know as 'English', 'Dutch' and 'Spanish' iris. Selections of hybrids of some of the species are forced and sold in winter in many different colours as cut flowers, or are grown as summer-flowering bulbs.

They were treated botanically by Dykes as a section of the genus *Iris*, but there are several other views about this. The Russian authority Rodionenko treats them as a separate genus, along with *Juno* (Subgenus Scorpiris) and *Iridodictyum* (recticulata irises) which are also given generic status. G. Lawrence regarded them as being a section, Xiphion, of the Subgenus Xiphium, which also included Section Reticulata. I prefer to accept the Spanish irises as a separate subgenus of *Iris* and not particularly closely allied to the reticulatas. Xiphium irises can be easily reognized since they are bulbous, the bulbs having papery, leathery or shell-like tunics, not netted as in the reticulata group. The roots are not swollen and fleshy as in the Scorpiris (juno) group, although in *I. boissieri* they are usually a little thickened. The leaves are channelled, carried at the base and up the stem. Unlike the Scorpiris the standards are held upright and are well-developed in all but *I. serotina*—in Scorpiris the standards are usually much reduced and held horizontally or are even reflexed in many cases. The falls have no beard, except in *I. boissieri* which has a narrow band of yellow hairs in the centre. The one to three flowers, produced on an unbranched stem 25–90cm tall, are normally produced in the wild between April and August.

Cultivation

On the whole the irises of this group are not difficult to grow, but some are rather tender in British gardens. In these cases it is best to plant them in a bulb frame or cold greenhouse. Pot cultivation is not very successful as they are too vigorous. They mostly grow on alkaline soils so it is best to include lime in any soil mixtures. All the species are best dried off in the dormant period in mid to late summer, but *I. latifolia* requires less drying than the rest. Further notes about cultivation are given under each species as they are described below.

Most of the species produce offset bulbs around the parent. These can be detached and grown on into larger bulbs, this process taking only one or two years. September is the best month for planting, just before growth begins and this is also the best time to lift established clumps and detach the young bulblets. The bulbs are best planted on a layer of sharp sand for extra drainage, and they should be not less than 10cm deep.

There are seven wild species with a number of varieties and numerous garden selections and hybrids many of which have been given cultivar names. A nurseryman's catalogue is the best place to search for these and I have not included a list here.

The Xiphium iris species

I. boissieri Henriques (Syn. *I. diversifolia* Merino). This differs from the rest in several ways. Its roots are somewhat fleshy, although not swollen as much as those of most of the juno iris. The stem is 30–40cm tall and the rather narrow leaves appear in the autumn. The solitary, deep violet-purple flower has a sparse yellow beard in the centre of the falls and the perianth tube is 3–5cm long. It flowers in June and occurs in northern Portugal and adjacent Spain, growing in black soil, rich in humus, among rocks at about 1500 metres. According to Dr Jorge Paiva the mountainous area in which *I. boissieri* grows is very cold in winter, often freezing and with snow almost every month at this time of year. It receives the highest rainfall in Portugal (1600–2800mm) and the temperature never rises above 20°C even in July, the hottest month. It seems therefore that *I. boissieri* might be fairly tolerant of outdoor cultivation in Britain.

I. diversifolia Merino. A synonym of *I. boissieri*.

I. filifolia Boiss. This grows 25–45cm in height with the leaves varying from almost thread-like (var. *filifolia*) to broadly linear (var. *latifolia*) and these appear in autumn. The one or two flowers are rich reddish-violet with a yellow stripe in the centre of the falls; the perianth tube is 1–2·5 (−3) cm long. It flowers in June and occurs in southern Spain, Gibraltar, Morocco and Tangier. This is a very beautiful species but is definitely tender and should be grown in a bulb frame or cold house. The long tube and richly coloured flowers distinguish it from any other species. There is however another method of identification, which is obvious even in the non-flowering plants. Major-General M. W. Prynne who is familiar with *I. filifolia* in the region of Gibraltar and southern Spain has noted that it possesses a purple- and white-blotched sheath which encloses the young shoot as it emerges from the soil. He gives its habitat as being in the *terra rossa* crevices between limestone rocks.

I. fontanesii Godron—See *I. tingitana*.

I. juncea Poiret (Syn. *I. imberbis* Asch. & Graebn.) A slender species about 30–40cm in height with very narrow leaves only 0·5–3mm wide, appearing in autumn. There are usually two scented flowers, bright yellow, with a perianth tube 3·5–5cm long. It flowers in June and occurs in southern Spain, Sicily and North Africa. I have not grown this species and cannot speak with any confidence on its cultivation. It is said to be rather tender and will therefore require a bulb frame or cold house. *I. juncea* can be recognized by its yellow flowers with a long slender tube, thus separating it from the yellow-flowered forms of *I. xiphium*. The bulb of *I. juncea* is covered by a rather tough shell-like tunic which splits at the apex into stiff points, unlike the more membranous tunics of other species.

Three varieties have been named: var. *mermieri* hort. ex Lynch, a sulphur coloured variety once offered by Krelage of Haarlem; var. *numidica* hort. from North Africa with paler flowers of a more lemon shade and var. *pallida* Barr ex Lynch, a large flowered variant in a soft canary yellow. This was at one time listed by Barr and Sons.

I. latifolia Miller (Syn. *I. xiphioides*). English iris. A strong-growing species usually up to 60cm in height with long leaves which do not appear until spring. They eventually grow up to 65cm long and are greyish-white on the upper surface. The flowers are large, usually two in each set of bracts and are violet-blue with a central patch of yellow on the falls. The perianth tube is about 0·5cm long. The falls have wide 'wings' on the lower portion or haft, a significant feature when comparing the species with *I. xiphium*. It flowers in June or July and occurs in north-west Spain and the French and Spanish Pyrenees in damp grassy places. This is an easy species to grow in the open border but the soil should not dry out too much.

Many garden forms of *I. latifolia* have been selected. The flowers are blue, violet, purple or white. However it is never yellow-flowered whereas *I. xiphium* does produce yellow forms. It is of course much better known in gardens as *I.*

xiphioides and it will be many years before Miller's correct name becomes accepted generally, if at all.

The English iris received its confusing name because Matthias de l'Obel saw plants of it in England, apparently near Bristol, in the 16th century and communicated the information to famous authors such as Dodoens and Clusius. Although it was soon known that *I. latifolia* was in fact a Pyrenean plant, the name 'English iris' remained.

I. lusitanica Ker-Gawl. See *I. xiphium*.

I. serotina Willk. This is perhaps the most distinct species of all the Xiphium group, very little-known until recent years and often thought to be a freak mutation of *I. xiphium*. It is clearly a distinct species, however. Thanks to Mrs Rosemary Strachey and Dr John Marr, I now grow this and it appears to be not difficult in cultivation. It is 40–60cm tall with narrow leaves which die away before flowering time. The stem leaves are short and bract-like, also becoming brown before the flowers open. The flowers are violet-blue with a yellow line in the centre of the falls and the standards are reduced to 1cm or less long and are so narrow as to be bristle-like, making this quite different from any other *Iris* of this group. The perianth tube is 0·5–1cm long. It flowers in August and inhabits south-east Spain in Jaen Province where it grows in the mountains in scrub. My plants have been grown successfully in a bulb frame and in deep pots in a cold greenhouse. They are only dried out for a short period in early autumn. The plants used to illustrate the *Botanical Magazine* plate number 733 were in fact taken from my garden on 14 August 1974.

I. serotina possibly has a wider distribution than that given above, for it is recorded by Maire in the *Flore de L'Afrique du Nord* as having been collected in the Middle Atlas and in the Rif Mountains. I have not seen the specimens of these and cannot therefore comment on their authenticity.

I. taitii Foster. See *I. xiphium*.

I. tingitana Boiss. & Reut. (Syn. *I. fontanesii* Godron). A robust species up to 60cm in height with long, arching, silvery-green leaves which appear in autumn. There are 1–3 large flowers from each set of bracts and they are pale to deep blue. The perianth tube is long and slender, usually 1–2·5cm (rarely to 3cm) long. This species flowers from February to May and occurs wild in Morocco and Algeria. It is a rather tender species in Britain and I have never succeeded with it outdoors in Surrey. Sir Frederick Stern, however, grew it extremely well in his hot sunny chalk garden in Sussex.

Var. *fontanesii* (Godron) Maire. This flowers rather later than most variants of *I. tingitana* and is a more slender plant with darker violet-blue flowers.

Var. *mellori* was described as a variety of *I. fontanesii* in 1973 by Collingwood Ingram. It is said to be very robust, up to 98cm tall, with purple flowers and is also reported as differing from *I. tingitana* in having a very rounded blade to the falls (rather pointed in *I. tingitana*) and from var. *fontanesii* in its flower colour and general robustness. I have not grown or even seen this plant and cannot comment on its relationships or cultivation. Captain Ingram recommends lifting the bulbs in summer and keeping them on a sunny shelf for three months before replanting. It grows in Morocco, in wet fields.

Hybrids between *I. tingitana* and *I. xiphium* have given rise to the race of garden plants known as Dutch iris.

I. xiphium Linn. Spanish iris. This is one of the most well-known of the group in cultivation and is often sold as a cut flower. It is a slender graceful species growing to about 40–60cm tall with the narrow leaves appearing in autumn. The one or two flowers are usually blue or violet in wild forms, with an orange or yellow median blotch on the falls, but they are sometimes wholly yellow or rarely white. The perianth tube is only 0·1–0·3cm long. In contrast to *I. latifolia*, the haft of the falls is narrow and unwinged. *I. xiphium* occurs wild in places which are marshy in spring but which dry out in summer. It is known to occur in Spain, south-west France, Corsica, southern Italy, Portugal, Morocco, Algeria and Tunisia. The flowering time is usually from April to May. Not surprisingly, with such a wide area of distribution there is a lot of variation and names have been attached to some of these variants as follows:

'Battandieri' (*I. battandieri* Foster). A white form with an orange ridge on the falls, from Morocco and Algeria.

'Lusitanica' (*I. lusitanica* Ker-Gawl.). The variant with yellow or bronze flowers from Portugal.

'Praecox' (var. *praecox* Dykes). A large-flowered variety, flowering in early April, rather earlier than most forms. It occurs in the Gibraltar area.

'Taitii' (*I. taitii* Foster). This flowers in late May and in gardens the name is usually applied to a form with clear pale blue flowers.

There are also many garden hybrids between *I. xiphium* and *I. tingitana*, possibly also with *I. latifolia* influence, and these are known collectively as the 'Dutch iris'. They have a very wide range of flower colour in yellow, bronze, pale blue to violet or deep mauve. They are often forced for the early cut-flower market.

Both wild forms and cultivars are easily grown in sunny, well-drained borders.

5   Subgenus Scorpiris (The juno irises)

This fascinating and attractive group of *Iris* species constitutes one of the larger natural groupings within the genus, with some 55 species at present recognized. Unfortunately they are not among the easiest of irises to cultivate although a few can be grown without protection in Britain.

Although they are, I believe, correctly treated as a subgenus and therefore under the Rules of Nomenclature take the name *Scorpiris*, I find it difficult to stop referring to them as 'the junos'. Since they are so well known in iris circles as this I will treat it as a sort of vernacular name without any nomenclatural implications!

The junos are bulbous plants, the bulb covered with papery tunics and consisting of a few fleshy scales attached to a basal plate which gives rise to

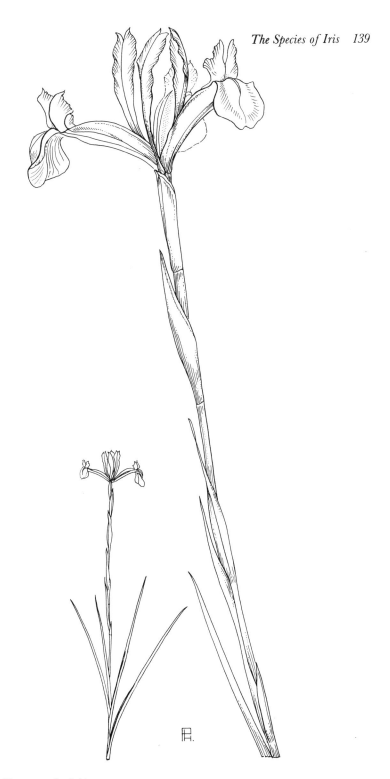

SUBGENUS XIPHIUM: *I. xiphium*

several thickened or fleshy roots, sometimes even radish-like in shape. These roots are replaced each year, new ones appearing during the flowering time and the old ones shrivelling away so that an undamaged dried bulb in its dormant state will have both old and new sets visible. It is necessary when handling the bulbs to treat them fairly carefully for the new roots are quite brittle and can easily become detached from the bulb. The channelled leaves are produced in one plane, that is to say they are distichously arranged, and vary enormously in size and texture from species to species. In the smaller ones such as *I. persica* there is very little leaf development and there may be only three or four, very short and produced all at ground level, with no stem and the flowers arising from the centre of the leaf cluster. At the other extreme, in *I. magnifica* for example, there is a long stem up to one metre in height with large leaves produced alternately all the way up it and carrying several flowers in the upper axils. The larger species resemble leeks or sweet corn in their general leaf growth habit. In *I. nicolai* and its relatives, the plant is short and stemless at flowering time but with broad leaves wrapped around each other to form a sort of tube through which the flowers appear. By fruiting time the leaves are fully developed and have been pushed well above ground in a cluster at the apex of a bare stem. In many species the leaves are grey-green, especially on the underside, and have a conspicuous transparent or white margin.

The flowers of this group of irises are fairly characteristic because in nearly all the species the standards are very much reduced, much smaller than the falls, and sometimes only visible as tiny bristles between the bases of the falls. They are often held out horizontally or deflexed so that the name 'standard' is hardly apt in the case of the junos. In only one species, *I. cycloglossa*, are the standards large and held in an erect or semi-erect position when the flower first opens. The falls are beardless but have a raised ridge in the centre and this can be undulate or sometimes crested like a cock's comb. In *I. sterophylla* subsp. *allisonii* there is a sparse beard of hairs on the lower part (the haft) of the falls, while in *I. tubergeniana* the ridge is crested to the extent of being almost beard-like.

The bracts enclosing the buds are rather useful in distinguishing between the species in this group, sometimes being green and rather rigid, sometimes transparent and papery, or occasionally somewhat swollen in the lower half and constricted at the apex so that they sheathe the perianth tube. The tube is normally quite long and may be as much as 8cm, depending upon the species. Going into finer detail one can find useful differences in the seed characters, some species having a whitish fleshy aril attached to the otherwise brown seeds. The pollen grains, seen under a powerful microscope are fascinating and diverse but it is beyond the purpose of this book to go into these features.

Distribution

The whole group of juno species is distributed mainly from central Turkey and the Caucasus, south to southern Israel and Jordan and east to northern Pakistan and the Pamir mountains in Russia Central Asia. One species only, *I. planifolia*, lies outside this area, occurring in Europe and North Africa.

Cultivation

Most of the junos are plants of the semi-arid steppe country and mountains of western and central Asia, receiving very cold snowy winters and hot summers with low humidity. Correspondingly they are on the whole not successful in the open ground in Britain for they either attempt to grow during muggy winter conditions and rot off or they do not receive the required warm dry summer dormant period. For most species, therefore, cultivation is a matter of providing some form of protection from the elements, either by growing them in deep pots (they have long and vigorous roots) in an unheated, well-ventilated greenhouse or in a bulb frame as described in the general cultivation chapter on page 7. At Kew, under the close scrutiny of Tony Hall, there has been great success using both methods and it is a joy to see large flowering clumps of such delightful species as *I. nicolai* and *I. albomarginata* in the bulb frames in the early spring. Unless there are comments to the contrary under any particular species the reader is to assume that I recommend cultivation by either of the above methods. A few species grow perfectly well in the open garden and exactly which will vary enormously depending upon the prevailing conditions in the country concerned, and on the local climate and soil of the garden. Obviously in parts of Europe and North America there are areas which experience temperature and rainfall patterns similar to those in the wild habitats of these plants so it will be up to the individual to experiment with outdoor cultivation. I can only speak from experience gained in southern England and can make no positive recommendations for other countries. Since most of the species grow naturally on limestone-derived soils, the medium in which they are grown should be well-drained and alkaline and I normally add garden lime for this purpose, but Michael Hoog recommends the use of dolomite chalk (ground magnesium and calcium carbonate) for this purpose. Extra feeding can be given in the growing season using liquid or solid fertilizers and as a personal preference I use a slow-release granular type placed on the surface of the soil in autumn or early spring so that it will be washed slowly down during watering. Watering should never be given so as to wet the foliage as it can be caught by the channelled leaves and held in the centre of the plant where rot may set in. Water is withheld altogether after the leaves have turned yellow, and the bulbs are then left dry until October.

I must also add some notes from the experiences of Michael Hoog in Holland, since they demonstrate how the behaviour can vary from country to country. He has found that a considerable number of the junos will grow outside in full sun in the sandy soil of Haarlem which dries out in summer. He recommends planting the bulbs shallowly, not more than 6–8cm deep and powdering them with dolomite chalk before covering with soil. In June or July when the foliage has wilted the bulbs are lifted and kept dry, at about 23°C, until November when they are replanted on newly prepared beds. In commercial production, at the nursery of Van Tubergen, he says that the bulbs are lifted every year except for seedlings which are left undisturbed for two years.

Propagation

Some species of juno increase naturally by the production of young bulbs alongside the parents, from lateral buds on the basal plate. In these cases propagation is merely a case of lifting and dividing the clumps during the late summer. It is also possible to propagate those species which are reluctant to divide up naturally by using the thickened fleshy roots. An inspection of the base of the bulbs will often reveal one or more of the small lateral buds. With a razor blade or scalpel it is sometimes possible to take off a bud, together with a small piece of basal plate and one of the fleshy roots, as a separate unit. The cut surfaces should all be dusted with a systemic fungicide before potting up the divisions in sandy compost. They should then be kept fairly dry until the normal watering season commences in autumn. I have succeeded with several species in this way but great care is needed or the parent bulb can be killed by rot setting in to the wounds.

Most species are best increased by seed if possible since this may well lead to an improvement, or at least maintenance, of vigour by singling out those seedlings which are most suited to the environment. Artificial pollination may be necessary and since many of the junos seem to be wholly or partially self-sterile it is best to cross pollinate between different plants of the same species, unless one is making an active attempt to produce hybrids in which case one would do experimental crossing between species. As a botanist I would in this case make a plea that the hybridizer records very carefully the details of parentage and publishes notes somewhere giving the origins of the offspring to avoid any future confusion. The seeds are sown in pots or prepared bulb beds and after germination are kept in growth as long as possible to build up reasonably sized bulbs before allowing them to go dormant in the normal way. Subsequently they can be treated as for fully grown bulbs.

Unfortunately a considerable number of species have never been in cultivation and many of those collected in the last few decades in Iran and Afghanistan appear to have been lost again. It is obviously very necessary if one collects or receives any interesting species to distribute them as widely as possible among keen iris growers in the hope that someone will be able to succeed with their cultivation.

The Scorpiris (juno) iris species

I. aitchisonii (Baker) Boiss. This is the most easterly occurring species in the group. It grows 15–50cm in height and has long, fleshy, but not swollen roots. The narrow leaves are well developed at flowering time, up to 40cm in length and only 4–8mm wide, spaced rather laxly up the stem. The stem is branched which is an unusual feature in the juno group. The one to three flowers have a perianth tube 2·5–3·5cm long and are either deep yellow or violet or sometimes bicoloured yellow with brownish or violet markings, especially on the small reflexed standards. The falls have a wide wing on the haft. In shape the standards are very variable from linear to oblanceolate or sometimes three-lobed. The flowering period is March–April. Its wild habitat is in Pakistan and eastern Afghanistan in moist grassy places which dry out in summer.

I have tried to grow this species, thanks to Dr Nasir of Rawalpindi who obtained seeds some years ago. Unfortunately it has not proved to be an easy plant. The seeds germinated well and the young bulbs have persisted but will not seem to increase to flowering size. In view of its damp wild habitat in spring perhaps one should try it out in the open ground rather than experimenting in pots and bulb frames.

I. alata Poiret. A synonym of *I. planifolia.*

I. albo-marginata R. C. Foster. (Syn. *I. caerulea*). This Russian species, originally described as *I. caerulea* by B. Fedtschenko, was given a new name by Foster because the epithet had already been used by Spach. As described by the *Flora of the USSR* it is rather a short plant with a stem 5–7cm in height, with long fleshy roots and short arched leaves, densely packed and concealing the stem. They are noted as having an abruptly narrowed apex and are 2–3cm wide at the base. The two to five flowers are blue with a tube about 4cm long and the haft of the falls has parallel margins, not widely winged. The blade of the falls is furnished with a white crest or ridge with a yellowish zone around it. The standards are oblanceolate and 2–2·5cm long. It flowers in June on stony slopes in the Tien Shan and Fergana mountain ranges of Central Asiatic Russia, at about 2000 metres.

The plants I have seen in flower in Britain do not exactly agree with the description given above, although I think this can be partly explained by changes in growth habit due to cultivation. The height is usually more like 20–30cm with the leaves spaced out rather than densely packed. It varies somewhat in shade of blue but the forms I have seen have all been remarkable for their intensity of colour and it is a really attractive and apparently fairly easy juno. At Kew it is grown very successfully in a bulb frame but it seems to accept pot cultivation equally well. One clone is particularly fine, with clear, almost 'electric' blue flowers produced freely. Individually the flowers of this species are not large but they more than make up for this by their vibrant colour.

Herbarium specimens I have seen indicate a height of only 5–8cm and show roots which are long and rather thin compared to those of most juno irises. The three to five leaves of this material show clear white margins and they are curved and only 1·2–1·5cm wide. The one or two flowers are about 4cm in diameter and the date of flowering, at variance with that given by the *Flora USSR*, is March or April. Clearly there is considerable variation in this species.

I. almaatensis Pavlov. A synonym of *I. subdecolorata.*

I. atropatana Grossheim. This is probably a synonym of *I. caucasica;* certainly it is very similar and the general description will be the same as for that species (see below). The difference is said to lie in the shape of the falls which in *I.*

atropatana have a narrowed portion (a sinus) between the blade and the haft. This however is a variable feature in *I. caucasica* and in this respect *I. atropatana* seems to fall well within the range of variability of the former. *I. atropatana* was first described in 1936 from specimens gathered in three places: Nakhichevan in southern Transcaucasia and at Oltu and Kagyzman in north-east Turkey. It is said to have yellow flowers.

I. aucheri (Baker) Sealy. (Syn. *I. fumosa* Boiss. & Hausskn. ex Boiss.; *I. sindjarensis* Boiss. & Hausskn.) A well known species, easy to grow in England. It is 15–40cm in height with fleshy, but not very swollen, roots. The eight to twelve leaves are well developed at flowering time, up to 25cm long and 2·5–4·5cm wide and they completely enclose the stem until fruiting time when the stem elongates and spaces out the leaves so that the internodes are visible. The three to six flowers have perianth tubes about 6cm long. The colour is very variable from blue to nearly white, with a yellow ridge on the falls, which have a broad wing on the haft. The standards are obovate, larger than in most junos, and are horizontal or deflexed. The flowering period is March–April. It occurs in northern Iraq, south-east Turkey, northern Syria and western Iran in rocky ground, on ledges or more rarely in fields, sometimes exceedingly abundant. There are also some herbarium specimens at Kew, apparently of this species, from Edom, Jordan, but the material is not of sufficiently good quality to be sure of its identity. Dinsmore attached the name '*I moabitica*' to the specimens from Qatrani in Moab, Jordan. Only fresh collections from the area will resolve the question as to whether they are *I. aucheri* or a distinct species.

Iris aucheri is not one of the most beautiful of the species in the group, having rather robust foliage for the size of the flower. Nevertheless it is possible to grow it in the open ground in southern Britain and any juno capable of being grown easily is a worthwhile plant to have. The colour form in commerce has flowers of a particularly fine blue and it is a strong grower, usually listed as *I. sindjarensis*. In the wild in south-east Turkey it varies enormously in colour from creamy-white to a deep violet and I hope that some of these forms will also find their way into more general cultivation.

I. baldschuanica B. Fedtschenko. This is known in England from one collection made in Afghanistan by Paul Furse in 1968, under the number 8206. It has persisted in cultivation for over 10 years thanks to the skill of Dr Jack Elliott, one of England's more successful growers of the juno group of irises. It grows 10–15cm in height and has very swollen roots. The leaves completely enclose the stem at flowering time and are slightly greyish-green. Later they are pushed up above ground level in a cluster at the apex of a short stem and reach 15–20cm long and 5–6cm wide when fully developed. The one to three flowers have perianth tubes 8–10cm long and are apparently variable in colour. The *Flora of the USSR* gives it as being yellow with violet veins on the falls whereas the plants collected by Paul Furse had creamy flowers and some pinkish-brown staining on the style branches and a yellow ridge on the falls. The margins of the haft of the falls fold downwards as in *I. nicolai* and *I.*

rosenbachiana and it is undoubtedly closely related to these two species. The standards are lanceolate, horizontal or slightly deflexed, and the style branches rather large and prominent. The flowering period is March–April. It grows in Russia in southern Tadjikistan and adjacent north-east Afghanistan, and in the latter area is recorded as being a plant of dry rocky slopes at 2100–2400 metres.

I. baldschuanica appears to be a relatively easily grown species and Dr Elliott has managed to propagate and distribute a small number of bulbs. It seems best in a deep pot in the alpine house or in a bulb frame.

I. bucharica Foster. (Syn. *I. orchioides* of gardens) This is a popular juno iris, for it can be grown outside in England with ease and does not require lifting during the summer. It is about 20–40cm in height with rather thin roots. The bright glossy green leaves are well-developed at flowering time and at this stage they are up to 20cm long and 2·5–3·5cm wide. Later the stem elongates and the leaves become spaced out up the stem. The two to six flowers produced in the upper leaf axils have perianth tubes 4·5–5cm long. The colour varies from golden yellow ('*I. orchioides*' of gardens) to nearly white with a yellow blade to the falls ('*I bucharica*' of gardens). On each side of the raised deep yellow ridge there are usually blotches or staining of green, brownish-green or dull violet. The margins of the haft of the falls are more or less parallel without a wide wing. The standards are deflexed or horizontal and are lanceolate or three-lobed. It flowers in April–May, and occurs in Russian Tadjikistan and north-east Afghanistan in stony places and edges of fields at 800–2500 metres. It is a widespread species in this area of Russia and the *Flora of the USSR* notes that it varies greatly in the form and size of the flower parts. Those from around Baldschuan, which is near the Afghan border, are noted as being particularly small-flowered. The deep yellow form has been grown for many years as *I. orchioides* but the Afghan collections of Paul and Polly Furse brought to light the fact that the species could vary in colour so much as to include this plant and the form grown as *I. bucharica* all in one wild population. It is now clear that the original plant named as *I. orchioides* by Carrière (see description below) is a different species, probably not in cultivation. It is said by Vvedensky to be pale yellow, changing to a pale violet shade as the flower dies and, more important, that the winged haft of the falls is 1·5–2·5cm wide. In *I. bucharica* and its forms the haft is unwinged and only 7–8mm wide.

From a garden viewpoint the wholly yellow form and the white and yellow bicoloured one are both worthy of cultivation as distinct plants and it would be preferable for convenience if cultivar names could be given to them so that they can in future be distinguished in catalogues and other literature.

Iris bucharica is one of the easiest members of this bulbous group to cultivate, requiring only good drainage and a sunny position. The name comes from Bokhara in Russia, the type locality.

I. cabulica Gilli. A little-known plant in cultivation and probably not grown at the present time although it has been collected several times during the last two decades. It is about 15cm tall with much-swollen roots and quite well developed strongly arching leaves which conceal the stem at flowering time. Eventually the leaves reach 25cm in length and about 2–4·5cm wide but stay at ground level, and are not raised up on a stem appreciably. The two to four flowers have a tube 5–8cm long and are slightly variable in colour from white to very pale lilac with a yellow ridge on the falls. The haft of the falls is not widely winged, but turns slightly downwards at the margins as in *I. nicolai*. The standards are reflexed and lanceolate. The flowering period is March–April and it grows in the Kabul area of Afghanistan on dryish slopes at 1800–2000 metres. It has been collected at Sher Darwasa, Munar Shakrie and Alishang, east-north-east of Kabul. In general appearance it is not unlike *I. baldschuanica* but the perianth tube is shorter and the margins of the haft of the falls do not turn down so markedly. Like *I. nicolai* and its relatives, *I. cabulica* possesses seeds which have a large white aril attached to one end.

I. caerulea B. Fedtschenko. A synonym of *I. albo-marginata*.

I. capnoides Vvedensky. This account is based on its description in 1971. It is said to have much-swollen roots and to have a stem about 8–15cm in height. The leaves sheathe the stem and are 0·8–1·6cm wide, curved and with a thickened margin. The one to three flowers are smoky grey-green with a yellow crest and they have a tube 4·5–5·5cm long. There is a wide wing on the haft of the falls, and the pale violet standards are three-lobed, about 1·3–1·7cm long. It was described from material collected on 10 May 1953 in the western Tian Shan and Ala Tau mountains near Tashkent and Kuramen.

I. carterorum Mathew & Wendelbo. This small species has been found only twice in eastern Afghanistan on dry slopes at 2200 metres. It is 7–10cm tall with narrow grey-green leaves and yellowish flowers spotted black on the lower parts of the falls and standards; the perianth tube is about 3cm long, the haft of the falls is unwinged and the bristle-like standards about 5mm long. It flowers in April or May.

I. caucasica Hoffmann. This is one of the easiest of the smaller junos to grow, but even so it requires the protection of a greenhouse or cold frame. It is usually about 15cm in height, with only slightly thickened roots. The leaves are usually five to seven in number and are well developed at flowering time, normally about 10–12cm long and up to 2cm broad, and rather greyish-green. At flowering time they sheathe and cover the stem but later the internodes become visible as the stem elongates. In this latter feature it is distinct from *I. pseudocaucasica*. There are one to four translucent green or yellowish-green flowers which have a yellow ridge on the falls and a perianth tube about 4cm long. On the haft of the falls there is a wider portion, or wing, which is however usually only slightly wider than the blade so that the overall shape of the fall is elliptical or nearly oblong. In some forms there is a slight constriction between the haft and the blade at the point where the blade folds downwards and it is this which was given as one of the distinguishing features of *I. atropatana* (see above). The standards are horizontal and oblanceolate. The flowering period is April–June and it occurs in central and eastern Turkey, north-eastern Iraq, north-west Iran and the Caucasus. The habitat is on well-drained slopes in limestone mountains at 1200–3500 metres.

There is a little difference between the plants from central and eastern Turkey and those from the Caucasus and the extreme north-east of Turkey. The latter plants, which are typical *I. caucasica*, have very ciliate margins to their leaves whereas those which are widespread in Turkey and which I have described as subsp. *turcicus* have glabrous-margined leaves.

The plants which occur around Tiflis, the type locality of *I. caucasica* subsp. *caucasica* are in general shorter and more compact than those from Turkey. Sir Michael Foster described a robust variety which he named var. *kharput* after the village, which is now called Harput near Elaziğ in east-central Turkey. Vigorous forms do occur in parts of the distribution of *I. caucasica* and there seems to be no reason why this particular collection should have a separate name, for the vigour is probably only related to local soil or climate conditions. The other features he mentions for var. *kharput* also seem to be within the scope of variation of the species. Grossheim described a var. *multiflora*, with up to 10 flowers from southern Transcaucasia and forma *caerulescens*, with pale bluish-violet flowers from Krachkessaman in Azerbaijan.

I. cycloglossa Wendelbo. The strangest juno of all, it looks more like a member of the xiphium iris group, although there is no doubt in my mind that it is placed in the right subgenus. It is a vigorous plant 40–50cm in height with long slender roots. The leaves are well-developed at flowering time, up to 30cm long and 1·5cm wide, scattered up the slender, sometimes branched stem. The one to three flowers are large for a juno, about 8–10cm diameter, with a tube 3·5–4·5cm long. The colour is a clear shade of blue-violet, the falls white towards the centre of the blade, with a yellow blotch, but no raised ridge. The claw of the falls is widely winged. Unlike any other juno, the large standards are about 4cm long, obovate and erect at first but falling outwards as the flower matures. The flowering period is in May and it occurs in south-west Afghanistan near Herat, in wet grassy places at 1450–1700 metres.

I. cycloglossa is a marvellous introduction and is now to be seen in a few specialist collections thanks to Hedge, Wendelbo and Eckberg who brought living bulbs from Afghanistan (no. 7727). It has proved easy to grow in England both in bulb frames and in the open garden. At Kew it has proved very successful planted into the raised bulb frame beds with other junos while in my own garden it has flowered well each year, first in a deep pot under glass and then planted out in a well-drained sunny place.

The large clear bluish-violet, clove-scented flowers are a delight and it is to be hoped that this is one juno which will become a well-known garden plant in the not too distant future.

The ecology of *I. cycloglossa* is rather extraordinary. It occurs only in a small area of north-west Afghanistan in wet ground near streams, unlike most other junos which grow in rather xerophytic conditions. The species has been collected only twice, as far as I know, to the south-west of Herat. The first gathering was made in 1949 by Koie and the second, exactly 20 years to the day later, was the collection from which the bulbs now in cultivation are derived.

I. dengerensis B. Fedtsch. A synonym of *I. narbutii*.

I. doabensis Mathew. This is yet another of the iris introductions of Paul Furse, who referred to it as 'Doab Gold' after the place it was collected and the flower colour. It grows about 10–15cm in height and has short, very swollen roots. The shiny green leaves completely conceal the stem and are well developed at flowering time, up to 20cm long and 4cm broad, so that it is a short, fairly leafy-looking plant. The one to five flowers have a tube about 7–8cm long. They are wholly deep yellow with a slightly darker crest on the falls, which have down-turned margins to the haft as in *I. nicolai, I. baldschuanica* and their relatives. The standards reflex slightly and are rather angular-ovate, less than 1cm long. In the wild it probably flowers in March or April. It has been collected only once as far as I know, by Paul and Polly Furse in the Hindu Kush mountains of north-east Afghanistan, about 100 kilometres west of the summit of the Shibar pass at 1700 metres. Any bulbs now in cultivation will be derived from this collection (P.F.5876) but it is undoubtedly very rare. At one time, it seems, only one bulb was in existence, in the expert care of Dr Jack Elliott. This flowered and was on loan to Kew for a while, when it was figured for the *Botanical Magazine* (tab.620, 1972). During this period I self-pollinated it and seeds were produced, these being distributed to various growers. *I. doabensis* is a delightful plant, for its flowers are not only a bright yellow but also have a fruity scent not unlike that of *Cytisus battandieri*.

I. drepanophylla Aitchison & Baker. This is one of the greenish-flowered species, about 10–30cm in height with short very swollen roots. At flowering time the leaves, which have a hairy margin, conceal the stem. Later they expand to up to 20cm long and 2·5cm wide, become strongly curved, and the stem elongates, spacing them out up the stem so that the internodes are visible. The two to eight flowers in the axils of the upper leaves are about 4–5cm in diameter and have a tube 3·5–4cm long. The colour varies a little from nearly yellow to greenish with a prominent deeper yellow crest on the falls. The haft of the falls has narrow down-turned margins and the standards are very narrow, almost bristle-like and slightly deflexed. It flowers in April in the Kopet Dag of north-east Iran and adjacent USSR and in north and north-west Afghanistan, usually on dryish slopes at 600–1700 metres in altitude.

   I. drepanophylla is rather similar in appearance to *I. kopetdagensis* but the latter has a more slender-looking flower which is usually greener and does not have the down-turned margins on the haft of the falls.

I. drepanophylla subsp. **chlorotica** Mathew & Wendelbo. This is a little-known juno from north-east Afghanistan, quite distinct in its distribution and as far as I know not in cultivation. The flowers are of a curious silvery green with a whitish or pale yellowish-green crest on the falls. It is known from only three collections, one of which was made by Paul Furse under the number 6505. This flowered once at Wisley and a colour photograph, under the Fursean name of 'Lime Green', can be seen in the *RHS Journal* of 1966, fig. 9. Unfortunately the bulb did not survive and there have been no other living introductions of this interesting plant.

I. edomensis Sealy. A curious juno which has probably only flowered in cultivation once and is now lost, it is a small species, growing to less than 10cm when in flower with no visible stem even at fruiting time. The four or five narrow leaves are very undulate on the conspicuously white margin and they also tend to be extremely falcate or coiled on the ground. The one or two flowers are about 4–5cm in diameter with a tube 4–6·5cm long and have a most extraordinary coloration. The falls are heavily stained, blotched and streaked an intense but dull purple, with a whitish ground colour and white border right round the edge. The standards, which are ascending rather than deflexed as in most species, are similarly coloured or creamy-white with only a few streaks and spots, and the style branches also vary in this way. On the haft of the falls there is a slight wing only, a little broader than the blade. The crest is yellow. The flowering period is January–February and the species seems to be confined to Jordan in the Edom region. It grows in the mountains around Petra and is a plant of rocky limestone or sandstone areas in dry *Artemisia* scrub at about 1300–1500 metres.

I. edomensis was first collected in fruit by Peter Davis in 1945 and living bulbs of this were flowered by Mrs Gwendolyn Anley in March 1946 in Woking, Surrey. Dr Davis's dried specimens and a flower from Mrs Anley's garden are represented at Kew. I have also been sent a specimen, collected in January 1976, by Dr Loutfy Boulos, the well-known Egyptian botanist who was at that time working in Jordan, and his excellent material is very similar to the original, suggesting that the species does not vary greatly in its curious colour combinations.

It will probably be a difficult species to grow since it occurs in areas which are very hot and dry in summer. The elongated bulb is covered by a mass of papery tunics which form a long neck reaching to the soil surface and I have noticed that other species with this character, such as *I. stocksii* and *I. postii*, are very tricky in cultivation. This feature seems to be the result of living naturally in an arid almost semi-desert habitat, far removed from the humid climate of the average Surrey garden!

I. eleonorae Holmboe. A synonym of *I. galatica*.

I. fosterana Aitchison & Baker. A very distinctive species with bicoloured flowers. The height is about 10–15cm and the long slender bulb often has many dull olive-coloured tunics; the roots are rather thin, not markedly swollen. The green, silver-edged leaves are well developed at flowering time and conceal the stem but later this elongates and they become spaced out and up to 17cm long and 4–8mm wide. The one or two flowers are about 4–5cm in diameter with a perianth tube 3·5–4cm long and have creamy-yellow falls and style branches contrasting with the sharply deflexed rich purple standards. There are sometimes brownish veins and a darker yellow blotch around the very prominent yellow crest. The style branches are large and nearly erect and the obovate standards, which are large for a juno at about 2–3cm long, are turned sharply down to the vertical position. *I. fosterana* flowers in March and April in north-east Iran, adjacent Turkmenistan and north-west Afghanis-

tan. It is thus mainly a plant of the Kopet Dag range where it grows in dryish
steppe country on stony slopes which are moist in spring but very dry in
summer. The altitude ranges between 750 and 2000 metres.

This striking species has been introduced many times into cultivation but is
still very rarely seen and appears to be difficult to keep for more than a year or
two. Sir Michael Foster, after whom it is named, noted in his *Bulbous Irises* of
1892 that it was not very amenable to cultivation and he expressed the doubt
that it would ever become common in our gardens. The best chance of success
seems to be in a cold greenhouse or a bulb frame where water can be withheld
completely in the summer. If it is pot-grown, the pots must be long ones to
accommodate the elongated bulbs and roots properly. When well suited,
I. fosterana has a clump-forming habit.

I. galatica Siehe (Syn. *I. purpurea* (Hort.) Siehe; *I. eleonorae* Holmboe). This is
one of the small Turkish junos, allied to *I. persica* and *I. caucasica*. It is 5–12cm in
height with slender fleshy roots and usually only three or four poorly
developed erect leaves at flowering time. Later these become arched and
expand to about 10–12cm long and 6–12mm wide at the base. The stem is
nearly absent and does not elongate in the fruiting stage. The one or two
flowers are very variable in colour from a rich reddish-purple throughout to
greenish-yellow with a purple blade to the falls or smoky silvery-purple with
darker falls. They are about 5·5cm in diameter and have a perianth tube
normally 4–6cm long. The crest is a prominent ridge in yellow or orange. Like
I. persica it has standards which are horizontal or slightly deflexed, extremely
varied in shape from spatulate to three-lobed, and 1–2cm long. In its
most usual forms, *I. galatica* has the bract and bracteole nearly equal, rather
rigid and straight or sometimes clasping the perianth tube at the apex. In
Turkey it flowers in March or April in rocky places in steppe country or in oak
and juniper scrub at 900–1700 metres altitude. It is a plant of the central-north
of Turkey and is known from the provinces of Amasya, Tokat, Sivas, Erzincan
and Erzurum as well as from the curious dry valleys around Nevşehir and
Kayseri, where it differs somewhat in appearance. It seems possible that in the
latter two places there are hybrid populations with *I. persica*, which is more
typically a plant of the southern lowlands of Turkey. The silvery ground-
colour of these plants is very reminiscent of that of *I. persica* and the bracteole is
slightly smaller and less stiff in texture, a character which is half way to the
papery white bracteole of *I. persica*. The geographical ranges of both species
could well meet in Cappadocia, but a lot of further study of these irises in the
wild is required before the true picture emerges.

This little juno is attractive in both its purple and greeny-yellow bicoloured
forms and in cultivation appears to be one of the easier species to grow. It is
certainly not so tricky as *I. persica* which rots off in the winter so easily in the
English climate. I have only grown *I. galatica* in pots under glass to date and it
persists with a reasonable amount of attention. The central Turkish plant with
silvery-purple flowers on the other hand seems to be as difficult as *I. persica*, as
one might expect from its habitat in places such as Göreme, where the

winters are cold and snowy but dry and the summers are scorching.

I. graeberana Van Tubergen ex Sealy. This is a robust and rather well-known juno, thanks to the firm of Van Tubergen who introduced and distributed it. Unfortunately the exact wild provenance is not known. It has a large bulb with long, not markedly swollen roots. The height at flowering time is about 15–35cm and the seven or eight broad leaves overlap and conceal the stem at this stage; later the stem elongates to up to 40cm and the leaves become spaced out with clearly visible internodes. The leaves are bright glossy green on the upper surface and greyish below, up to 15cm long and about 1·5–3·5cm wide, and have a whitish margin. In vigorous specimens there are four to six flowers which are about 6–7·5cm diameter with a perianth tube 5–6cm long. The predominant colour is blue, perhaps with a little violet or lavender suffusion, especially on the standards and styles, while around the white undulate crest is a pale, almost white, strongly veined area. The rounded apex of the falls is a deeper, almost cobalt blue while the haft is paler and has wide wings. The standards are obovate, about 2–2·5cm long, with a long-pointed tip. Its behaviour in the wild is unknown but in Britain it flowers in March or April.

I. graeberana is named after Paul Graeber who collected it in central asiatic Russia. The exact locality is not recorded and the best information available seems to be 'Turkestan' which covers a wide area! In general appearance it resembles *I. aucheri* which, however, is a plant of much farther to the west, in western Iran, Turkey, Iraq and Syria. Its nearest relative appears to be *I. willmottiana* which was also described from Turkestan. This however differs in having less wide wings to the hafts of the falls—about 1·5cm wide in the latter and 2–2·5cm wide in *I. graeberana*. It seems quite probable that the two are variants of one species but until a thorough study of the range of the variation can be made in the wild it is best to leave things as they are. At present both species are little-known and the observations on *I. graeberana* are all based on the one introduction by Graeber in the early part of the century.

In cultivation it is not a difficult species and can apparently be cultivated in the open ground like *I. bucharica* and *I. magnifica* although I have no personal experience of this. It is reported that seeds are freely produced in cultivation.

I. heldreichii Siehe. A synonym of *I. stenophylla*.

I. hippolyti Vvedensky. This account is based on Russian literature. In the original description *I. hippolyti* is given as a short plant with a stem only 10cm long from a bulb with much-swollen roots. The leaves are held close together, sheathing the stem, and are 1–1·5cm wide. The solitary flower is pale violet with a yellow area in the centre of the blade of the falls and the crest or ridge is white and undissected. The haft of the falls has a wing, the tube is about 4cm long and the standards are 1·5cm long. The information supplied is that it was collected in April 1934 in rocky places near Koktscha in the Kyzyl Kum desert. It is compared with *I. willmottiana* but is said to differ in having narrower leaves and a yellow blotch in the centre of the falls. I once flowered a similar juno, sent to me by Frank Kalich of Albuquerque, New Mexico. This originated in the Kara Kum desert which is adjacent to the Kyzl Kum and it is possible that this was in fact *I. hippolyti*, although in a few details they do not agree. Such is the variation in most junos, however, that these small discrepancies might mean very little.

I. hissarica O. Fedtsch. A synonym of *I. narbutii*.

I. hymenospatha Mathew & Wendelbo (Syn. *I. persica* var. *isaacsonii*). This is one of the relatives of *I. persica* but is rather distinct in its leaves, bracts and flower colour. It grows about 7–12cm in height and has a bulb with a long papery neck and rather thin, only slightly fleshy roots. The three or four narrow very long-tapering leaves are rather short and only 4–9mm wide at flowering time, increasing somewhat in size towards the fruiting stage. They are dark green above with a silvery-white margin and greyish below with very prominent silvery veins. There are one to three nearly stemless flowers with perianth tubes about 7cm long, and they are nearly white with violet veins; on the falls there is a scarcely raised yellow crest surrounded by a distinct bluish-violet zone. Like the rest of the *I. persica* group, this has winged falls, the winged part about 2·5cm wide. The standards are also whitish, oblanceolate and horizontal, 1·5–2cm long and 0·5cm wide, narrowed to a slender claw.

It is the bract and bracteole which distinguish it very clearly from *I. persica*, for they are both very thin and soft in texture, and silvery-transparent. In *I. persica* the bract is rather stiffly erect and green and only the bracteole is thinner and semi-transparent. In flower colour *I. hymenospatha* is much whiter in appearance than most forms of *I. persica*, and it lacks the deep coloured blotch at the apex of the falls which is a noticeable feature of *I. persica*.

I. hymenospatha occurs in southern Iran, particularly in the Shiraz region where it grows on stony hillsides of limestone, in open situations or in scrub, at 1500–2000 metres altitude. The flowering time is between January and March. It has proved to be a difficult plant to grow in Britain. The best chance of success is to grow it in a bulb frame with very good drainage, plenty of air in the autumn and winter and a good long dry period in the dormant season.

There is a tendency for the more northerly-occurring *I. pseudocaucasica* to approach *I. hymenospatha* in its characters in the region just to the south of the Elburz mountains and some comments about such a collection made by John Ingham will be found under *I. pseudocaucasica*. *I. hymenospatha* is distinguished from *I. caucasica* and *I. pseudocaucasica* by its much longer perianth tube.

I. hymenospatha subsp. *leptoneura* Mathew & Wendelbo. This is a distinct variant of the species which occupies a geographical range rather more to the north west of Iran, around Hamadan and Kermanshah, and crossing into eastern Iraq. It has wider leaves, as much as 1·6cm in some specimens, and these tend to be rather more strongly curved and do not have the very prominent nerves on the underside which are so characteristic of subsp. *hymenospatha*.

I have not had the opportunity to grow this plant, and although it was collected several times by Paul Furse in the 1960s it appears to be not in cultivation. It is probably easier than subsp. *hymenospatha* since it occurs in a slightly less sunscorched region at varying altitudes up to 2250 metres. It flowers on the whole a little later, between February and May.

I. inconspicua Vvedensky. This is described as having slender roots and a short stem, only about 5cm long, and closely arranged curved leaves so that the internodes are not visible. The one to three flowers are pale lilac with dull green spots on the blade of the falls and the crest is white and dissected. The haft of the falls is not winged and is only 6–7mm wide with more or less parallel margins. The perianth tube is 4·5cm long and the standards about 1cm, trilobed and pointed at the apex. It was described from material collected in the western Tian Shan mountains. I have seen no herbarium specimens or living plants and cannot do more than repeat the details given in the original description of 1971.

I. kopetdagensis (Vved.) Mathew & Wendelbo. This varies in stature from 10–35cm in height and has bulbs with much-swollen roots. The leaves are well developed and conceal the stem but usually become spaced out by the fruiting stage. On average they are about 10–30cm long and 1–2·5cm wide and strongly curved. The three to nine flowers (rarely solitary) in the upper leaf axils are about 4–5cm in diameter with a tube 4–5cm long and are green with a yellow crest on the falls. The haft of the falls is very narrow with upturned margins so that it is gutter-shaped and not at all winged. The standards are slightly deflexed, very much reduced so that they are bristle-like, only 5–10mm long. Flowering in the wild is March to May and it occurs on dryish stony slopes or edges of fields at 1000–3000 metres altitude. As the name shows it is found in the Kopet Dag range of north-east Iran and adjacent Russia and extends into north-west and central Afghanistan.

   I. kopetdagensis is most like *I. drepanophylla* in general appearance but can be distinguished by the upturned margins of the haft. In the latter they are wider and down-turned. The margins of the leaves is finely hairy in the latter and more or less glabrous in *I. kopetdagensis*.

I. kuschakewiczii B. Fedtschenko. A striking Russian species little known in cultivation in the west, it is in the hands of a few specialists growers. It is a stocky species 10–15cm in height when in flower with four to five leaves completely concealing the short stem. They are deep green with a very distinct white margin and are about 1–1·5cm wide. The one to four flowers have a tube 3·5–4·5cm long and are about 6·5–8cm in diameter. The overall colour is a very pale violet but on the falls there are some blotches and interrupted lines in very dark violet, on either side of the prominent white crest. On the haft is a slightly widened portion or wing, about 1cm across at its widest point. The standards are 1–1·5cm long, more or less three-lobed or toothed. In Russian Central Asia it flowers in April or May on stony slopes in the northern foothills of the Tien Shan mountains.

   In cultivation in Europe, *I. kuschakewiczii* does not seem to be a difficult species and would probably take to gardens where junos such as *I. magnifica* and *I. bucharica* thrive. Unfortunately there has never been enough material to experiment with and one is loath to plant the only bulb outside in case the theory is wrong! Maurice Boussard in Verdun has been quite successful with it but I must admit to being less skilful with its culture in Surrey and my one bulb has dwindled in size.

I. leptorhiza Vvedensky. This is unknown to me and there appears to be no herbarium material available in Britain. The description is therefore taken from literature sources. The stem is said to be absent and the three to four leaves densely packed together so that the total height cannot be more than about 10–15cm. The bulb has thin roots, not even fleshy, so this feature makes it a little uncharacteristic of the juno group, and the bulb coats are described as being brown with strongly prominent ribs. The leaves are strongly curved and about 5–10mm wide at the base. It appears that there is probably only one herbarium specimen in existence and this has a solitary flower, described as being violet-green by the collector. The perianth tube is 4cm long and the dimensions of the falls suggest that it is quite a small flower, perhaps about 4cm across. The haft of the falls is unwinged, about 5mm wide with nearly parallel margins, and the crest is raised but undissected. The standards are narrowly lanceolate, only 3mm wide and 1cm long. The one specimen known was collected in March in the Pamir-Alai region of Soviet Central Asia by Michelson, growing in the mountains of Tabakcha.

I. linifolia (Regel) O. Fedtschenko. I have not seen this species in the living or dried states and can only construct a description from other sources. The bulb has short much-swollen roots and papery tunics. It appears to have a stem only 5–10cm in height but at the same time the curved narrow leaves (4–7mm wide) are spaced out so that the internodes are visible. The one, or rarely two flowers are quite small, with a perianth tube about 4cm long and are described as being pale yellow with a darker blade to the falls and a whitish crest which is toothed or sometimes dissected. The haft of the falls is about 6mm wide and unwinged with nearly parallel margins, and the standards are more or less three-lobed and about 1cm long.

   I. linifolia flowers in May and June at about 2500 metres on rocky slopes of the Tien Shan and Pamir-Alai mountains. It is sometimes cited for the Hindu Kush range in Afghanistan but there seems to be no real evidence for this.

I. magnifica Vvedensky. A well-known species in cultivation, often obtainable from nurseries and easy to grow without protection. It is a robust plant 30–60cm in height with many, 3–5cm wide, shiny green leaves scattered up the stems so that the internodes are visible at flowering time. The three to seven large flowers are up to 7·5cm in diameter and are pale lilac with a yellow zone surrounding the undissected whitish crest. The perianth tube is about 4·5–5cm long and a prominent feature of the falls is the wide wings on the haft, usually about 2–2·5cm at the widest point. The standards are 2–3cm long and obovate. In the wild it flowers in April and May in rocky places in the mountains of Central Asia especially around Samarkand.

I. maracandica Vvedensky. A little-known species described as having a stem 10–15cm tall, hidden by the well-developed leaves which are 1·5–2cm wide. The one to four flowers have a tube 3–4·5cm long and are yellow with a widely winged haft to the falls. It occurs wild in the stoney foothills of the Pamir-Alai mountains of Central Asia, flowering in March or April.

I. microglossa Wendelbo. A very distinct species introduced into cultivation in Britain in the 1960s by Paul Furse under the Fursean vernacular name of 'Salang Blue'. It grows 10–40cm in height and the bulbs have long, thickened roots. The leaves are well-developed at flowering time, not completely sheathing the stem so that some internodes are visible, and are up to 25cm long and 1·5–2·5cm wide. They are distinctly bluish or greyish-green on the upper surface and finely hairy on the margin. There are one to four flowers, about 4·5–5·5cm in diameter, pale lavender blue to nearly white with a white or very pale yellow crest. The perianth tube is 3–4·5cm long and the haft of the falls is broadly winged. The horizontal standards are oblanceolate and about 1·5–2cm long. *I. microglossa* grows in north-east Afghanistan in Kataghan province where it grows on dryish slopes at 1700–3000 metres. It is found especially on the Salang Pass and has been noted by several collectors in that region, in addition to Furse.

I. müllendorfiana Bornmüller. This is a name written on a herbarium specimen of *I. galatica* at Kew. Apparently however it is not a validly published name.

I. narbutii O. Fedtschenko (Syn. *I. dengerensis* B. Fedtsch.; *I. hissarica* O. Fedtsch.) Not a well-known species in cultivation by any means, but it has been seen once or twice and I have made up the description from literature. In flower it is about 10–15cm in height with four to six curved, dark shiny green leaves only 4–6mm wide, concealing the stem but sometimes becoming spaced out a little in the fruiting stage. The bulb is distinctive, being long and slender with tough tunics reaching to the soil surface, and the roots are not very swollen. The one or two (rarely more) flowers are about 5cm in diameter and are bicoloured, usually pale violet with a dark violet velvety blotch at the tip of the falls and with a raised whitish crest surrounded by a yellow zone. The standards are deflexed and are bright violet and large for a juno, being obovate and about 2·5–3·5cm long. The perianth tube is 4–5cm long and the haft of the falls is unwinged, about 4–7mm wide. This is probably the best colour form, for it is a variable plant and the background colour can be a dirty yellowish colour and the blade of the falls sometimes has the violet patch only in the centre with a whitish margin.

It is a central Asiatic species, growing on the rocky or clayey foothills of the Pamir-Alai and the Syr Darya mountains. It is known from the environs of Samarkand and Tashkent.

I. narynensis O. Fedtschenko. I have not seen material of this Russian species, but from its description it is a small plant with much-swollen roots and a stem 5cm long, hidden by the falcate leaves which are 5–10mm wide. The one or two flowers are pale violet with a darker blade to the falls which have an unwinged haft and a white ridge; the tube is 4·5–5cm long. It grows in the

Tien Shan range on stoney slopes in the Naryn river valley, flowering in March.

I. nicolai Vvedensky. It is probable that this should be combined with *I. rosenbachiana* as a colour form of that species. However, since it is well known in cultivation and the range of variation in wild populations is not well known, matters should perhaps rest until more information is available. For *Flora Iranica* (1976) however Per Wendelbo and I decided to treat them as one under the older epithet *I. rosenbachiana*. Perhaps some system of varietal names is the answer.

In the colour form grown as *I. nicolai* this is one of the most striking of junos, but unfortunately it is a little temperamental. It grows 12–15cm in height and has very swollen, radish-like roots. The leaves are poorly developed at flowering time and completely enclose the stem in a sort of tubular sheath. By the fruiting stage however, they develop to 25cm long and 5–6cm broad and the stem elongates to 5–10cm, carrying the leaves in a cluster at its apex rather than spacing them out up its length. The one to three flowers are about 5–6cm in diameter and have an enormously long perianth tube, usually 8–11cm, and the haft of the falls has down-folded margins. The colour is rather bizarre in the form normally seen in cultivation; in overall ground colour it is whitish or pale lilac, but the falls have a velvet-like deep violet apex and two similarly coloured nerves on each side of the prominent rich orange crest. The rest of the flower is coloured pale bluish-lilac or nearly white with a stain of dull purple on the style branches. The standards are obovate but usually the margins are in-rolled so that they look narrower, and they are about 2–2·5cm long. It flowers in February to April on clayey hill slopes in north-eastern Afghanistan (Kataghan province) and in the adjacent Pamir-Alai mountains of Russia, usually at altitudes of 1000–2000 metres.

I. nicolai is a superb plant for a bulb frame or alpine house and appears to be reasonably easy if given plenty of light and air during the growing season. It can vanish almost overnight if any overhead water lodges in the leaf bases so that watering must always be carried out with the utmost care.

As known in cultivation at present, *I. nicolai* is the name given to the colour form described above and *I. rosenbachiana* is a reddish-purple-flowered plant with more leaf showing at flowering time. There are various reasons however to suppose that there are more variations than this. Dykes, in his *Handbook of Garden Irises*, described the latter as being dark crimson-purple on the blade of the falls and the rest of the flower white or faint purple. This appears to be somewhat intermediate. It is also on record, in *The Garden* of 16 June 1888, that it is found wild 'in two varieties, both growing together, the flowers of one form being blue and those of the other of a fine violet'. Sir Michael Foster, writing in the *Gardener's Chronicle* of 1889 says of *I. rosenbachiana* 'I find that hardly any two plants are exactly alike'. Foster also mentions a soft creamy yellow form sent to him by Regel—possibly closer to the plant now known as *I. baldschuanica*. Regel described a variety *albo-violacea* of *I. rosenbachiana* in which the whole flower was nearly pure white except for a large blotch of deep violet on the blade, with touches of violet and a sheen of rich yellow.

The amount of leaf development also varies considerably and, judging by the herbarium material at Kew, it is possible to find plants with scarcely any leaf at flowering time, through to forms in which the leaves are well developed, some with near-white flowers and others with purplish flowers.

So it would appear that we are probably dealing with one variable species, two very different clones of which have become well known in cultivation giving us the impression of distinct species.

I. nusairiensis Mouterde. A comparatively recently described species from Syria, related to *I. aucheri*. It is a stocky plant with thickened, but not swollen, roots and a 7–10cm stem completely sheathed within the five or six broad (2–3cm) leaves which are usually well-developed and somewhat curved. They are bright green on the upper surface and greyish below and the top-most leaves have a swollen, paler whitish-green base which is veined darker. This feature it shares with *I. aucheri*. The one or two large flowers are about 6–7cm in diameter and are pale blue or whitish with a creamy or very pale yellow ridge or crest. There is usually some darker spotting or veining on the widely winged haft. The segments usually have wavy edges which give the flowers a rather 'frilly' appearance. The perianth tube is 4–5cm long and the sheathing bracts which tightly enfold the tube are covered with a very grey 'bloom'. The standards, as in *I. aucheri* are quite large for a juno, being about 1·5–2cm long and broadly oblanceolate. It flowers in April in rocky places in Syria and is named after the Jebel Nusairi.

I. odontostyla Mathew & Wendelbo. A little-known species, collected only twice by Paul and Polly Furse and now probably lost to cultivation. It grows about 13cm in height and has long-necked bulbs with a cluster of long slender rather wiry roots, not at all swollen. The mass of papery tunics reach to the soil surface and in this feature it is similar to *I. postii* and *I. stocksii*, both semi-desert species. The four or five leaves conceal the very short stem and are well-developed at flowering time being up to 18cm long and about 1·5cm wide at the base, and they have a white margin. The solitary flower is about 5–5·5cm in diameter with a perianth tube 4cm long and there are wide wings on the haft of the falls. The colour is predominantly greyish or silvery-violet with an orange-yellow crest surrounded by a near-white area. The name is taken from the very toothed lobes to the style branches, a rather prominent feature of the flower. It grows on steep stony slopes and rock ledges at about 1200–1500 metres in the Herat region of Afghanistan where it is subjected to very arid conditions for much of the year.

This species was collected under the numbers Furse 8853 and 8857 and has flowered in cultivation in Britain. The toothed style branches, greyish-purple colour, rounded blade to the falls and wiry roots make it distinguishable from *I. stocksii* and *I. platyptera*, also from Afghanistan.

I. orchioides Carrière (not *I. orchioides* of gardens, which is a form of *I. bucharica*). This is a name frequently seen in literature and in gardens but it appears that there has been some confusion and the true claimant to the name is a rather different plant from central Asia which is very rare or unknown in cultivation in the west. It grows about 20–35cm in height when in flower and has fleshy, but not swollen, roots and five to seven leaves which just conceal the stem when the first flowers open. The stem elongates rapidly, however, and the internodes are visible by the late flowering stage. The light green leaves are straight or curved and vary in width a great deal, usually about 1–3cm broad and up to 18cm long at flowering time. The flowers, usually three or four, are about 5cm in diameter and pale yellow with a slight suffusion of pale purple, especially as they become older. This colour, and the fact that the haft of the falls is widely winged, clearly separates it from '*I. orchioides*' of gardens. The 'wing' is about 2cm wide and the perianth tube 3–6cm long, usually stained with violet. In the centre of the blade of the falls, surrounding the dark yellow raised and toothed crest is a deeper yellow zone sometimes broken by violet-green lines. The standards are 8–15mm long and narrowly linear or more or less three-lobed. *I. orchioides* occurs in the Syr Darya and Tien Shan mountains at about 1500–2000 metres where it flowers between March and May. It has been recorded from the Tashkent and Ugam areas and the *Flora of the USSR* notes that it is extremely variable.

My notes on this species have been made from literature and from herbarium specimens at Kew, for I have not grown the true plant. The original description and illustration are not precise enough to be absolutely certain about the characters of the species. It is quite clear, however, that the plant described as *I. orchioides* by Vvedensky in the *Flora USSR*, and the dried specimens sent to Kew, represent a different species from that which is often grown under the name in Britain. The latter is quite clearly a yellow form of *I. bucharica* and further comments under that species can be found above.

I. palaestina (Bak.) Boiss. This is very similar to *I. planifolia* with which it shares the common feature of having spiny pollen grains. In overall appearance it is very similar and although it is often a translucent greenish colour, it is also reported that it can have whitish or almost bluish flowers in which case the association with *I. planifolia* would be even more marked. There is a slight difference in the shape of the stigma, it being bilobed in *I. planifolia* but not in *I. palaestina*. It is a native of the eastern Mediterranean at low altitudes in Israel, the Lebanon coast and probably also southern Syria. It often inhabits stony coastal areas and olive groves where it flowers in December to February.

I. palaestina is an easy species to grow but like *I. planifolia* is very susceptible to virus diseases. It is also a little tender and if pot-grown, the pots should be plunged in sand to avoid freezing. I have only grown the green-flowered form which is a delightful plant.

I. palaestina var. **caerulea** Post. A synonym of *I. postii*.

I. parvula Vvedensky. A very distinct-looking little plant, unfortunately not in cultivation as far as I know. It is about 10–12cm tall with a slender bulb, which has very swollen roots. The three to four straight leaves are spaced out on the stem so that the internodes are clearly visible, and they are strap-like and only

5–10mm wide, abruptly narrowed to a short point at the apex. The one or two small flowers are about 4cm in diameter with a perianth tube 4cm long and their colour is described as being pale yellowish-green with dirty yellow spots and veins at the side of the pale yellowish-green dissected crest. The haft of the falls is unwinged, about 5mm wide and the standards are very narrow, almost bristle-like, and about 5mm long. It is a native of central Asia in the Hissar and Zeravshan mountain ranges where it flowers in May and June at 2500–3000 metres on rocky clayey slopes.

I have not seen this interesting little plant in the living stage, but herbarium specimens show that it is very distinct with its short, parallel-sided leaves spaced up the stem. Vvedensky notes that in one area specimens with tapering leaves occur and in another, plants with apparently violet flowers, although he admits that further studies into the species are required.

I. persica Linn. This has the distinction of being the first plant to be illustrated in the *Botanical Magazine*, in 1787. In spite of its name it is not known to occur in Iran. In the 18th century it was considered to be such an easy plant to grow that beds of it were planted outside, and bulbs were also placed in jars of water on windowsills for their early flowers in much the same way as we treat hyacinths now.

I. persica is a dwarf plant about 10cm in height, with rather thin, not markedly swollen roots. The three to four long, tapering leaves are up to 10cm long and about 5–15mm wide at the base, green on the upper surface and greyish below, with a white or translucent margin. They elongate very little after the flowering stage and the short stem remains below ground so that at fruiting time the capsules are held only just above the surface. The one to four flowers are about 5–6cm in diameter and have a long tube usually 6–8cm in length. The colour varies considerably and names have been given to some of the forms, as mentioned below. Usually the overall appearance is a pale dull translucent silvery-grey, sandy-yellow, brownish or grey-green colour with a darker purplish or brownish blade to the falls. The crest is yellow and very prominent, often spotted brown or purple. The standards are horizontal or deflexed and vary greatly so that they can be oblanceolate to more or less three-lobed, and about 1·5–2·5cm long. The haft of the falls is very widely winged as in the rest of the *I. persica* group of species from western Asia.

In the wild it flowers between February and April in open stony places or in sparse scrub or pine woods. It is typically a plant of southern and south-eastern Turkey, keeping to the lower and drier foothills between 100 and 1650 metres. It is very common in the eastern Taurus and Amanus mountains and extends eastwards nearly to Lake Van and southwards into northern Syria and northern Iraq.

From the related *I. galatica* and *I. stenophylla*, which also occur in Turkey, it can be distinguished not only on its semi-transparent flower colour but also on the texture of the bract and bracteole (or 'spathe valves' as they are sometimes called). In *I. persica* the bract is rigidly erect and green while the bracteole is often whitish, smaller and more like tissue paper. In the other two species the bract and bracteole are more equal and both green.

I. persica is a well-known species and has been a popular garden plant for centuries. Unfortunately it is not easy to keep for long periods and it is likely that the large plantings of the 18th century relied on new importations of bulbs every year. In Holland it seems to thrive on the dune sand and I remember being shown by Michael Hoog rows of it in flower on a miserable day in February when nearly horizontal rain was lashing across the bulb fields. The Dutch bulbs are lifted for drying each summer, which is certainly a necessary practice in countries where summers can be as wet as the winters. If grown in pots or in a bulb frame the moisture control is much easier of course. My own experience of *I. persica* is that it likes an alkaline soil, plenty of water during the spring and a long dry warm dormant period. Water must never be given from above so as to wet the leaves, unless the frame or alpine house is extremely well ventilated.

As mentioned above, various varieties of *I. persica* have been described, and these have sometimes been regarded as separate species. Those referable to *I. persica* are as follows:

I. bolleana Siehe. The flower is plain yellow with a violet blotch on the falls. It comes from the Taurus mountains, in the low limestone foothills.

I. haussknechtii Siehe (= *I. sieheana* Lynch and *I. persica* var. *magna* Hort.)

The flowers are large, silvery-grey, marked with purple-red on the falls. It is also from the Taurus and the Amanus mountains, in *Pinus brutia* woods. *I. persica* var. *mardinensis* from further east has a similar colouration.

I. issica Siehe. The flowers have a straw-coloured ground colour. It is described as occurring in the mountains by Issus near Adana.

Other names which have been linked with *I. persica* but which belong to other species are:

I. eleonorae Holmboe—see *I. galatica*.

I. galatica Siehe—see under species of that name.

I. heldreichii Siehe—see *I. stenophylla*.

I. persica var *issaacsonii* Foster—see *I. hymenospatha*.

I. persica var. *purpurea* Hort. (*I. purpurea* (Hort.) Siehe)—see *I. galatica*.

I. stenophylla Hausskn. & Siehe ex Bak.—see under species of that name.

I. tauri Siehe ex Mallet—see *I. stenophylla*.

I. planifolia (Miller) Fiori & Paol. (Syn. *I. alata* Poir.; *I. scorpioides* Desf.) A well-known species, long cultivated in Britain and notable for being the only wild juno species in Europe. It grows 10–15cm in height and has a large bulb with fleshy, not swollen roots. The numerous arching shiny green leaves are 1–3cm wide and well-developed at flowering time and they are densely arranged, concealing the stem which elongates only a little in the fruiting stage. The one to three flowers are about 6–7cm diameter and are normally bluish-violet with darker veining around the yellow crest. Occasionally white forms occur in the wild. The perianth tube varies from 8–15cm long and there is a wing on each side of the haft of the falls. The standards are held out horizontally and are about 2–2·5cm long, oblanceolate and usually toothed. It occurs in rocky places up to 300 metres in Mediterranean regions, where it flowers from November to February. It has been recorded in southern Spain and Portugal, Sardinia, Sicily, Crete, Morocco, Algeria and Libya.

This, and the very similar *I. palaestina* from the eastern Mediterranean, are very distinct from other junos in that they have pollen grains covered with minute spines.

In the past *I. planifolia* has been imported into the horticultural trade in large quantities from wild populations and in some of its areas has become very rare. There are recent records of it being sighted in considerable quantities in other places, but this does not mean that it is fair game for the plant collector! It is unfortunate that, although not difficult to grow in a bulb frame or alpine house, *I. planifolia* contracts virus diseases very easily, much more readily than other junos in my experience. The leaves become streaked with paler green and loss of vigour results.

I. platyptera Mathew & Wendelbo. A rather dull species, probably not now in cultivation. It was illustrated in the *RHS Journal* of 1966, fig. 8 under the name 'Old Smokey' which was how Paul and Polly Furse saw its colours when they collected it. It grows 8–14cm in height and the bulbs have short, swollen roots. The four to six grey-green leaves are well developed at flowering time and they conceal the very short stem which does not elongate in the fruiting stage. The leaves are about 10–18cm long and 1–3cm wide, strongly curved and with a very distinct white margin. As mentioned above, the flowers, two or three in number, are a rather smokey colour and somewhat transluscent, about 5–5·5cm in diameter. The overall appearance is dirty purple or brownish violet with a yellow crest. The haft of the falls is widely winged, the perianth tube about 4cm long and the standards 0·8–1·2cm long. The latter are oblanceolate and often three-lobed, the central lobe being long and pointed and the lateral lobes rounded. It flowers in March and April on dry stony slopes at 1800–2700 metres altitude and occurs in eastern Afghanistan, especially in the Kabul region, east to the Pakistan border area where it has been recorded in the Kurram valley.

Although *I. platyptera* (the name refers to the wide wings of the haft) has been collected many times there have been few living introductions for the purposes of cultivation. The Furse gatherings of the early 1960s are, it would seem, now lost.

I. popovii Vvedensky. This is one of the *I. nicolai-I. rosenbachiana* group and one I have not seen. The general description is the same as that given for *I. nicolai* (see above) but the leaves are described as being almost fully developed at flowering time. The two to four flowers are blue or light violet, possibly with a yellowish blade to the falls, but even the Russian *Floras* are not very sure of the true flower colour. The difference between it and the other members of its group, given by the *Flora USSR*, is that the standards are only 4mm wide whereas in *I. baldschuanica, I. nicolai* and *I. rosenbachiana* they are 5–12mm. At the bottom end of this scale there would be only 1mm difference which seems rather trivial when one remembers that in the junos the standards are usually extremely variable in shape and dimensions. *I. popovii* was collected in June by melting snow on clayey soil at 3600 metres in the Pamir-Alai mountains. The Garden-i-Kaftar pass is the type locality.

I cannot comment on the status of this as I have no knowledge of it other than that gleaned from Russian sources.

I. porphyrochrysa Wendelbo. Yet another species which was brought into cultivation by Paul and Polly Furse but probably now lost again. It is a small plant, only about 10cm in height and the bulbs have long, thickened roots. The three to five leaves are well developed at flowering time, about 10–15cm long and 5–10mm wide, rather grey-green and concealing the stem even in the fruiting stage. The one to three flowers are about 4·5–5·5cm in diameter with a tube 3·5–4cm long and are of a distinctive colour, being almost bronze except for a deep yellow blade to the falls, which have an orange crest. The haft of the falls is narrow and unwinged, and the bronze standards are reflexed and almost bristle-like, or sometimes trilobed. It flowers in May or June on dryish slopes amid spiny cushion plants such as *Astragalus* and *Acantholimon* at 2700–3000 metres in central Afghanistan.

This is Paul Furse's 'Shibar Bronze' or 'Shibar Gold' juno which is illustrated in the *RHS Journal* no. 91, fig. 12 (1966).

I. postii Mouterde (Syn. *I. palaestina* var. *caerulea*). A species little-known in cultivation but in the wild very common over a wide area and frequently collected as herbarium material, it is about 8–20cm in height and has a very long-necked bulb covered with a mass of papery tunics, and many thin wiry roots. The three to six leaves are well developed at flowering time, up to 16cm long and 0·8–1·5cm wide. They are prominently white-margined, have a long-tapering apex and are arched or coiled. The stem is concealed at flowering time but by the fruiting stage the internodes are visible between the leaves. The one to three flowers have a tube 2·5–4·5cm long and are about 5·5–6cm in diameter. Although the overall colour is violet it is blotched and veined dark violet on a pale ground of lavender. On either side of the yellow crest is a white band. The haft of the falls is widely winged and the standards are horizontal or slightly inclined and 1·2–1·8cm long. *I. postii* is a plant of sandy and gravelly soils in semi-desert and fields at 300–600 metres where it flowers in February to April. It is not uncommon in the Jazira and western desert areas of Iraq, in eastern Syria and eastern Jordan. I have also seen specimens collected in irrigated flax fields near Baghdad. My attempts at cultivation have been very unsuccessful and I decline to offer advice!

I. pseudocaucasica Grossheim. A common species in the wild, it is often called *I. caucasica* in literature; but the two seem to be quite distinct in the shape of the falls. It is a small plant, about 6–20cm in height with fleshy, but not swollen, roots. The leaves, usually three or four in number, are almost fully developed at flowering time, being 5–18cm long and up to 3cm wide at the base. They are strongly arched and prominently veined with a white, usually rough margin and they completely enclose the stem at flowering time. By the fruiting stage however they are a little more spaced out but the internodes are seldom visible.

The one to four flowers are 5–6cm in diameter with a tube 3–4cm long. In colour they vary considerably but are usually a translucent yellowish-green or icy blue with a yellow crest on the falls, and these are widely winged on the haft. The lanceolate standards are horizontal, very variable in shape and 1·2–2·2cm long. It is a native of northern Iran, north-east Iraq, south-east Turkey and southern Transcaucasia. It flowers in April to June and inhabits mountain screes and stony hillsides at 600–3500 metres altitude.

This small species is particularly common in the Elburz mountains where it can form quite large clumps on the upper rocky slopes. Some of the populations which occur south of the Elburz approach *I. hymenospatha* in their leaf characters. Such an example is one I have grown which was collected by John Ingham on the Tehran to Ghom road at 1350 metres. The leaves are long and tapering and very prominently silver-veined like those of the latter species but the bracts are green, not silvery and tissue-like as in *I. hymenospatha*. This particular collection seems to do well in cultivation, which *I. hymenospatha* certainly does not.

In Turkey, *I. pseudocaucasica* has been collected only once or twice in the mountains south and south-east of Lake Van. Only the yellow form has been recorded and the plants are very dwarf.

I. pseudocaucasica is distinguished from *I. caucasica* in having a very wide wing on the haft, narrowing abruptly, whereas in *I. caucasica* the haft is only slightly wider than the blade, giving an overall elliptical shape to the falls. In general terms, *I. caucasica* elongates its stem rather more after flowering than does *I. pseudocaucasica* so that the leaves become more spaced out. Furthermore the latter usually has only three to four (rarely five to six) leaves whereas *I. caucasica* normally produces six to seven leaves (rarely four).

I. purpurea (Hort.) Siehe. A synonym of *I. galatica*.

I. regis-uzziae Feinbrun. A newly described species which has been introduced to Britain and has flowered at Kew in December 1979. The height is about 10cm when in flower and the bulb has fleshy, not swollen, roots. The five to seven broad leaves are densely packed at the base and conceal the short stem and are curved, well developed at flowering time and about 2·5–4cm wide. There are one or two flowers about 5–6cm in diameter and these can be pale blue, lilac or translucent greenish yellow. The perianth tube is about 4cm long. On the haft of the falls there is a wide wing, and on the blade a yellow median ridge. The standards are 2–2·5cm long, horizontal and rather spoon-shaped with a short point. In the wild it flowers in January to March in calcareous rocky places at 500–1000 metres altitude. According to Prof. N. Feinbrun it has been recorded in Israel in the highlands of the central Negev desert where the annual rainfall is 100–200mm. Material has also been collected in southern Jordan around Ras el-Naqb by D. Birkinshaw and by Prof. Loutfy Boulos. I became aware of the Jordan plant in 1977 when some bulbs of the Boulos collection flowered in Lund, Sweden, with Dr Mats Gustafsson. He sent me colour photographs, and Prof. Boulos sent the herbarium material and more colour slides. It was clear that this was an unknown species and I fortunately corresponded with Prof. Feinbrun before giving it a name, for she was already in the process of describing *I. regis-uzziae* from the Negev region of Israel, having known of its existence there for many years. We came to the conclusion that the Jordanian and Israeli plants in fact belonged to the same species.

I. regis-uzziae is named after Uzziah, King of Judah for 52 years in the first part of the eighth century BC

In cultivation at Kew, a bulb sent by Prof. Feinbrun flowered in December in a frost-free greenhouse. In this form the flowers were a translucent greyish-blue with a yellow ridge on the falls and some dots and streaks of purplish-brown along the centre line of the haft. There was a faint scent, contrary to the observations of Prof. Feinbrun who reports it to be unscented. My own bulb of this same collection has not flowered but has grown strongly in an unheated greenhouse.

I. rosenbachiana Regel. This comparatively well-known juno was figured over 100 years ago in the *Botanical Magazine* and has been around in cultivation for at least this length of time. It most resembles *I. nicolai* in flower shape and habit and indeed the two are sometimes regarded as colour forms of one species, united under the earliest name of *I. rosenbachiana*. For the general description, and notes about the variability of *I. nicolai* and *I. rosenbachiana* lengthy comments were given under the former species and need not be repeated here.

The plant we know in cultivation as *I. rosenbachiana* has rich deep purple flowers with a bright orange crest on the falls which of course makes it quite different as a garden plant to the white and violet bicoloured *I. nicolai* clone. The leaves are a bright shiny green and they are moderately well-developed at flowering time with outward curving tips whereas in *I. nicolai* they are poorly developed, incurved at the tips and of a more greyish-green.

In the wild, *I. rosenbachiana* flowers in March or April on stony slopes up to 2000 metres altitude. It occurs in central Asia in the Pamir-Alai range. Although plants looking like *I. nicolai* in colour have been collected by Paul Furse in northern Afghanistan, no purple forms comparable with *I. rosenbachiana* have been recorded outside Russia. One dried specimen at Kew, from Darwas, shows quite considerable variation in flower colour from purple to near white, and a variable amount of leaf development so that it seems quite probable that it is correct to merge the two under one name. Only studies of wild populations can resolve the matter satisfactorily.

In cultivation, *I. rosenbachiana* is not a difficult plant in a bulb frame or alpine house, but overhead watering must be avoided as the broad leaves catch and trap the water and can result in rot setting into the crown.

I. schischkinii Grossheim. These remarks are based on only a photograph of the type specimen of this species. It resembles *I. caucasica* very closely. It is a very robust plant, 20–30cm in height with several large flowers, probably about 6cm in diameter, with all the parts rather larger than those of the average *I. caucasica*. The haft of the falls is wide, but only slightly wider than the lamina, which is an acceptable shape for the very variable *I. caucasica*. The other details of the plant seem to be rather similar also. It was described from Russia near Lake Batabad, above the town of Bitschenach in the Nakhichevan area, flowering at 2100 metres in May 1947.

I. scorpioides Desf. A synonym of *I. planifolia*.

I. sindjarensis. Boiss & Hausskn. A synonym of *I. aucheri*.

I. stenophylla. Hausskn. & Siehe ex Baker. (Syn. *I. tauri* Siehe ex Mallet; *I. heldreichii* Siehe) A member of the *I. persica* group, but recognizable as a distinct species and occupying a slightly different geographical area and habitat in Turkey. It is a dwarf species, about 6–12cm in height, and the bulb has fleshy roots, not swollen. The four to five leaves are short at flowering time, expanding later to about 10–20cm long and 0·5–1cm wide, curved, usually green above and below and without a prominent white margin. Sometimes they are slightly grey-green below. The stem is very short and concealed by the leaves and does not elongate in the fruiting stage so that the leaves remain in a cluster at ground level. The normally solitary fragrant flower is about 5·5–6·5cm diameter with a tube 6–9cm long and has widely spreading falls, usually almost horizontally. The markings and colour vary somewhat but the overall impression is violet-blue or lilac-blue, often with an even darker blade to the falls. Around the prominent yellow crest is a whitish zone spotted with violet. On the haft of the falls is a wide wing and the horizontal to deflexed standards are 1–2·5cm long and vary from spoon-shaped to three-lobed. It occurs mainly in the Cilician Taurus mountains in southern Turkey and has been recorded from Isparta province eastwards to the high Bolkar Dağ at the eastern end of the Taurus. In the wild it grows on open rocky slopes and alpine pastures at 400–2000 metres and flowers in March to May, usually soon after the snow melts.

I. stenophylla is a delightful plant with its richly coloured flowers and it appears to be reasonably easy to grow—much more so than *I. persica*. The colour is variable and this factor resulted in the appearance of more than one name for the plant. Sometimes the whole flower is a strong violet-blue and in other forms the ground colour is pale lilac-blue with the darker violet confined to the tip of the falls accompanied by heavy blotching around the crest. From *I. persica* it can be distinguished on the flower colour, by the more numerous green leaves without a white margin and on its bract characters. The bract and bracteole in *I. persica* are unequal in size and/or texture whereas in *I. stenophylla* they are both green and usually closely sheathe the tube.

I. heldreichii and *I. stenophylla* are names referring to the same plant. Siehe wrote, in the *Gardener's Chronicle* of May 1901, '. . . *I. heldreichii*, which has already been described as *I. stenophylla* (a name which, for various reasons, I do not recognise)'. Unfortunately he did not give his reasons, but *I. stenophylla* is the older name and therefore takes priority. *I. tauri* was thought to be different but this view cannot reasonably be supported and it must be seen only as one of the colour variants of the species. In fact in the Kew herbarium there is a specimen of *I. tauri* labelled 'came amongst *I. heldreichii* from Siehe' indicating

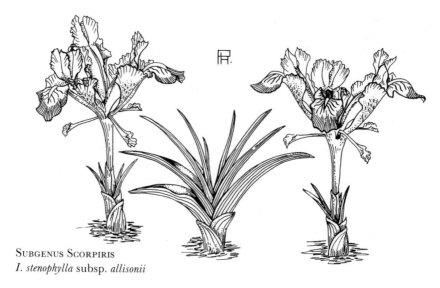

SUBGENUS SCORPIRIS
I. stenophylla subsp. *allisonii*

that mixed populations of the colour forms occur. J. D. Hooker in the *Botanical Magazine* of 1901 observes that 'a plant of *I. tauri* was received at the Royal Gardens, Kew in 1900, from Mr Siehe, in a batch of bulbs of *I. stenophylla*.' It seems that the darker forms were being selected out as *I. tauri* and those with a paler ground colour as *I. stenophylla* or *I. heldreichii*. George Mallet in *The Garden* of 1902, writes of *I. tauri* that he could distinguish eight colour forms in the batch of bulbs he had, so it is obviously very variable although always in the violet-blue area of the spectrum.

I. stenophylla subsp. ***allisonii*** Mathew. This is a distinct variant which has apparently been collected only three times in the western Taurus. The first gathering was made in 1975 by myself, Turhan Baytop and Chris Brickell when some non-flowering plants were found growing with *I. pamphylica*. Realizing that the locality was far west of that of any known juno I asked John Allison and Peter Ball to search for it during their visit to an area slightly to the east in 1977. They also collected non-flowering material and the bulbs from this collection subsequently flowered in cultivation. I returned in 1979 to the first locality but it was again out of flower. Although it bears a general resemblance to *I. stenophylla* subsp. *stenophylla*, there are some distinguishing features, and its ecology appears to be slightly different. In subsp. *allisonii* the leaves are rather numerous for such a small plant and I have counted between six and ten from flowering-sized bulbs. They are often very undulate on the margins and are broad (1·5–1·8cm) when mature. The flowers have a shorter tube, about 5·5cm, than in subsp. *stenophylla* and they are generally of a bluer tone with very prominent darker spotting and streaking on the falls. The Allison & Ball collection has flowers of a clear paler blue than those of my 1979 collection (number 9636). A further interesting feature of subsp. *allisonii* is a line of hairs along the middle of the haft of the falls, absent in subsp. *stenophylla*. This interesting plant is fairly easy to grow in a bulb frame or alpine house in southern England, possibly because it originates from a relatively high rainfall area of Turkey. The known area of distribution lies to the east of Antalya in the mountainous region formerly known as Pamphylia. It grows in scree at the edge of pine woods at 850–1500 metres altitude.

I. stocksii (Bak.) Boiss. This is a juno which will probably never persist in British collections for more than a year or so at a time since it is a semi-desert plant. It grows 10–20cm in height and the bulb has long slender fleshy roots which are not markedly swollen. The bulb tunics are very papery and persistent forming a long neck reaching the surface of the soil. The leaves are well developed at flowering time, about 10–15cm long and 0·5–1·5cm wide, concealing the stem at first but becoming spaced out by the fruiting stage on the elongated stem. The one to four flowers are about 5·5cm in diameter with a tube 3–5cm long, and are lavender or bluish-violet with a yellow crest on the falls. The haft of the falls is widely winged and the 1–2cm long, obovate standards are held horizontally. *I. stocksii* flowers in March or April and grows at 1000–2700 metres on dry stony hills which are very hot in summer. It is a native of central and south Afghanistan and western Pakistan, especially in the Quetta region, northwards to Kabul.

The species has rarely been in cultivation and although in recent years Paul Furse introduced some more bulbs those have not apparently persisted.

I. subdecolorata Vvedensky (Syn. *I. almaatensis* N. Pavlov) This is probably not cultivated in the west at present. I have seen only one dried specimen and the following notes are from this and Russian literature. It grows only 5–8cm in height when in flower and the slender clump-forming bulbs have tough chestnut-brown tunics and thin wiry roots. The four to five leaves are well developed at flowering time but even so are only 6–7cm long, and are curled back so that they touch the ground. They have a very white margin. There is no elongation of the stem after flowering so the plant remains compact at ground level. The one to three flowers are only 3–3·5cm in diameter and have a tube about 4·5cm long. The colour is described as being pale translucent dirty greenish or lilac-tinged, with two dirty violet veins on the falls and a dull green blotch surrounding the white, dissected, almost beard-like crest. The falls are not winged on the haft and the horizontal standards are 5–7mm long and narrowly oblong, angular or three-lobed. It flowers in March or April and grows on clayey hills in the foothills of the Syr Darya range near Tashkent.

I. svetlanae Vvedensky. I have not grown this species and can only repeat the information given in Russian literature. It is described as having much-swollen roots, a stem 10–15cm in height and closely-packed leaves which are curved and white-margined, about 3–4cm wide. The one or two large flowers are a rich yellow with green veins on the falls and the crest is deep yellow and untoothed. There is a wide wing on the haft of the falls and the perianth tube is 4·5–5cm long. The standards are 1·2–1·6cm long, rhomboidal in shape and pointed. It was collected in gypsum hills, flowering in March, in the valley of the Kaschka-Darya in the Pamir-Alai mountains of Central Asia. It is named after Svetlana Kovalevskaja, a friend and colleague of Vvedensky who was accompanying him when *I. svetlanae* was collected in 1958.

I. tadshikorum Vvedensky. This is a plant I have tried to grow from Russian seeds, but the bulbs perished before reaching flowering size. My notes are therefore again gleaned from Russian descriptions. It has much-swollen roots and a stem about 5cm in height which is completely sheathed by the 6–10mm wide leaves. These are curved, narrowed gradually to the apex and have a prominent margin. The two to four flowers (rarely solitary) are thought to be pale violet with a white, dissected crest. The perianth tube is 3·5–4cm long and the haft of the falls is unwinged with almost parallel margins. It inhabits the Pamir-Alai mountains around Darwaz in Russian Central Asia, occurring on stony slopes at about 2500 metres where it flowers in June.

I. tauri Siehe ex Mallet. A synonym of *I. stenophylla*.

I. tubergeniana Foster. It is very unfortunate that this very distinctive and attractive iris seems to be lost to cultivation now, at least in Britain. I hope that this comment will result in a letter stating the contrary! It is a stocky plant only 10–15cm in height with long, slightly thickened roots and four to six broad white-margined leaves closely packed together and concealing the stem. The widest is up to 2·5cm broad and they are curved and almost fully developed at flowering time, so it is rather a leafy-looking plant with the flowers nestling in the strong leaves. The one or two flowers are about 5–6cm in diameter with a tube 4·5–5cm long and are yellow with a fimbriate crest, appearing almost beard-like. On either side of the crest are a few greenish-violet veins and dots. The haft of the falls is only narrowly winged or almost unwinged and the 1·5cm long, three-lobed standards are deflexed. It flowers in March or April and occurs wild in Russian Central Asia. Although the first collections were without precise localities, Vvedensky states that he has seen it growing in the Syr Darya foothills in clayey soil. I have seen a dried specimen collected near Darbaza (? Darwaz) on hills of stony clay, and a colour slide of the plant in cultivation, but have not had the pleasure of growing it myself. Certainly it was known as a living plant fairly recently in Britain, for it was grown by Paul Furse and E. B. Anderson, but I can find no one who has bulbs at present. It received an Award of Merit in 1901 and used to be grown outside in Britain in warm, sunny, well-drained places.

I. vicaria Vvedensky. This species has been confused with *I. magnifica* but the two are apparently distinct. It grows about 20–50cm in height from a very large bulb which has thickened but not swollen roots. There are five to seven arching leaves which are well-developed at flowering time, about 15cm long and 1·5–3cm wide, and well spaced up the stem with clearly visible internodes. The flowers usually number two to five but occasionally there are more or less than this depending upon the vigour of the plant. They are 4–5cm in diameter and have a tube 4–4·5cm long. The colour is pale bluish-violet with darker lines on the unwinged haft and a yellow blotch around the yellow or whitish crest, which is undulate or crisped. The standards are about 1·5–2·5cm long, oblanceolate to three-lobed and pale violet with darker veins. It flowers in March or April and inhabits crevices and rocky slopes at about 1000–1500 metres in Russian Central Asia. There are records from the Pamir Alai range and I have seen specimens from the Hissar mountains above Schargun, which is the type locality. Vvedensky also gives for its distribution the Baisun mountains, and the Chulbair area.

I. vicaria differs most obviously from *I. magnifica* in having no wing on the haft of the falls. It is in cultivation in Britain but in only one or two specialist collections.

I. vvedenskyi Nevski. It is very appropriate that one juno should be named after the botanist who did so much work on the group. Unfortunately it appears to be unknown in cultivation. It is a small plant up to 10cm in height when in flower with the bulb having short swollen roots. The few, curved leaves are slightly spaced out on the stem and are only 4–5mm wide. There is usually just one flower which is pale yellow with an orange, toothed crest. Since the falls are described as being only 2·5–3cm long and the tube a similar length, it must be a rather small flower. The haft of the falls is narrow and unwinged and the standards are about 6mm long, varying from linear to three-lobed. It was described from Kμh-i-Tang in the Pamir-Alai mountains, flowering in May.

I. warleyensis Foster. A vigorous plant, rare in cultivation but apparently not a difficult juno. It grows 20–45cm in height from a bulb with only slightly thickened roots. The six to seven curved green leaves have an inconspicuous margin and are well developed at flowering time, up to 20cm long and 1·5–3cm wide, and are spaced out on the stem to give clearly visible internodes. The one to five flowers are 5–7cm in diameter and have a tube 4·5–5cm long. The colour varies somewhat from light to deep violet or purplish-blue, the blade of the falls margined with white and with a yellow zone around the whitish to yellow, toothed or crisped crest. The haft of the falls is unwinged and the blade turns vertically downwards giving a rather characteristic appearance. The deflexed standards are 1–2cm long and vary from narrowly linear to more or less three-lobed. It flowers in March or April on the lower stony slopes of the western Pamir-Alai mountains. I have seen dried specimens from the mountains around Samarkand, but it is said to grow near Derbent and Bokhara also.

I. warleyensis was introduced by an expedition to Central Asia organized by Mr J. Hoog of Haarlem and was described in 1902 by Foster after Miss Ellen Willmott's famous garden, Warley Place. It is a colourful species with its contrasting violet and yellow falls and my experience is that it is best grown, like *I. bucharica*, as a border plant among other herbaceous plants.

I. wendelboi Grey-Wilson & Mathew. This was named after Per Wendelbo, a good friend and colleague who has done so much work on the flora, especially the bulbous plants, of Iran and Afghanistan. It is a dwarf species about 10cm in height with short, swollen roots. The three or four leaves are well-developed at flowering time, very grey-green, about 15–20cm long and less than 1cm wide, and are strongly arched or coiled on the ground. They conceal the short stem but by fruiting time this elongates and the internodes are sometimes visible. The one or two flowers are about 4·5–5·5cm in diameter and are deep violet with a bright golden-yellow frilled crest along the centre of the unwinged falls. The tube is short, about 3·2cm long and the very narrow standards only 5mm long. It flowers in March or April in dry sandy hills at about 1700 metres in south-west Afghanistan. Christopher Grey-Wilson and Tom Hewer collected this under numbers 561 and 575 and it is to be hoped that a few bulbs are still in cultivation in Britain in specialist collections.

I. willmottiana Foster. Another fine species, named in 1901 for Miss Ellen Willmott of Warley Place in Essex. It is a robust but short plant about 15–25cm in height with thickened but not swollen roots. The short stem is completely covered at flowering time by about eight broad shiny green leaves densely packed together at first but becoming less so towards fruiting time as the stem elongates. The four to six large flowers are about 6–7cm in diameter and are a lovely soft lavender or pale purple colour with blotches of white mingled with marks of a deeper lavender on the blade of the falls. The tube is about 5cm long. Although the haft of the falls is winged, about 1·5cm wide, the transition from claw to blade is gradual—not abrupt as in, for example *I. persica* or *I. magnifica*. The overall shape of the fall is therefore nearly elliptical. The crest is whitish, not markedly crinkled, and the standards are about 1·5cm long, varying from nearly diamond-shaped to three-lobed. It grows wild in Russian Central Asia near Derbent in the Pamir-Alai mountains where it flowers in May.

I. willmottiana was found in 1899 by a collector for the firm of Van Tubergen who later sent bulbs to Sir Michael Foster. I have not seen it in cultivation in Britain and it seems highly probable that it has been lost. The plant grown as *I. willmottiana* 'Alba' seems to me to be one of the many forms of *I. bucharica*.

I. xanthochlora Wendelbo. This is yet another species first collected and introduced by Paul Furse but probably not now in cultivation. It is 10–15cm tall and has much-swollen roots. The three to five well-developed arching leaves are up to 20cm long and 2cm wide at flowering time and completely cover the stem so that there are no visible internodes. The one to three flowers are greenish-yellow with a deep yellow crest on the falls which are not winged on the lower part. The tube is 5–6cm long. The standards spread horizontally and are 1–1·5cm long and narrowly oblanceolate. It flowers in May to July on dryish mountain slopes at about 2500–3500 metres in north-eastern Afghanistan.

I. xanthochlora is very like *I. kopetdagensis* from north-western Afghanistan and Iran. The main difference lies in the slightly smaller flowers of *I. xanthochlora* in which the haft of the falls is tapered from base to apex. In *I. kopetdagensis* the haft has more or less parallel margins.

I. zaprjagajewii N. Abramov. A beautiful and interesting species named as recently as 1971. It is related to *I. nicolai* in having a bulb with almost radish-like roots, short leaves at flowering time and flowers with the haft of the falls down-turned at the margins. It grows only 10–15cm in height and the flowers appear at the same time as the broad (up to 4cm) grey-green leaves which are clustered at the base of the short stem, completely sheathing it. The one to three flowers are about 5·5cm in diameter with a tube 6–9cm long. They are entirely white, except for the yellow crest on the falls, and sometimes a little bluish suffusion towards the base of the segments and on the tube. The standards reflex slightly and are about 1cm long, more or less rhomboidal or oblanceolate in shape. It occurs near the town of Nischup at 2200 metres in the south-western Pamir mountains of Russia. In the wild it flowers in April or May but in Britain somewhat earlier than this.

I was very fortunate to be sent a few bulbs of this species soon after it was described. It did grow well and flower for a year or two before disappearing and I am happy to have actually seen this beautiful species in the living state rather than as a squashed dried specimen!

I. zenaidae Vvedensky. This is named after Zenaida Botschantzeva who studied the cytology of many of the Russian juno irises. It is described as having much-swollen roots, curved leaves about 2cm wide, which are closely packed together at first, but eventually spaced out up the stem, and one to three large flowers. These are violet-blue, strongly spotted violet on the haft of the falls and have a white or violet crest. The tube is 4·5–5·5cm long and the haft of the falls is widely winged. The standards are broadly oblong and 2–2·8cm wide. It occurs wild in the Fergana region of the Tien Shan mountains.

I. *zenaidae* is said to differ from *I. magnifica* in its darker, heavily spotted flower colour.

Hybrids Hybridization in the junos has fortunately not 'caught on' in the same way as in many iris groups. It would perhaps be a pity if it did, for there are so many beautiful wild species already it seems unnecessary. The only case for such action would be to provide easily grown species for general garden use, using those species which are already known to be tolerant of the British climate, such as *I. magnifica* and *I. bucharica*.

There are a number of hybrids in cultivation, for example *I.* 'Warlsind'.[*I. warleyensis* × *I. aucheri* (= *I. sindjarensis*)] *I.* 'Sindpur' [*I. aucheri* (= *I. sindjarensis*) × *I. galatica* (= *I. purpurea*)].

Natural hybrids are also on record and Vvedensky has noted possible crosses in the wild between *I. narbuti* and *I. maracandica*, *I. orchioides* and *I. narbuti*, *I. subdecolorata* and *I. narbuti* and *I. vicaria* with *I. bucharica*. In Turkey I have seen *I. caucasica* plants which may have had some influence of *I. persica* and in central Turkey the populations of *I. galatica* are in need of further study for they also may be hybridizing with *I. persica*.

6   Subgenus Hermodactyloides (The 'reticulata irises')

This group contains the well-known dwarf bulbous irises which are most useful for rock gardens or cold greenhouse displays in the early spring. There are about 10 species, distributed from central and southern Turkey to the Caucasus, Iran and Russian Central Asia, and southwards to Israel. They were treated as Section Reticulata by Dykes and as a separate genus, *Iridodictyum* by Rodionenko.

The group is easily recognized by having strongly netted tunics covering the bulbs and usually one or two long basal leaves which are roughly square in cross-sectional shape. Exceptions to this are *I. bakerana*, which has nearly cylindrical leaves, and *I. kolpakowskiana* in which there are three or four narrow channelled leaves. The flowers are of typical iris form except in *I. danfordiae* where the standards are reduced to tiny bristle-like organs between the falls. All the reticulatas are dwarf plants, at most about 10–15cm high when in flower. In fruit they have nearly stemless erect capsules, except in *I. pamphylica* where they are pushed up well above ground on a stem and are pendulous.

Cultivation

Iris reticulata and its cultivars, *I. histrioides* and *I. winowgradowii* all grow very well when planted in the open ground, requiring merely a sunny, reasonably well-drained position. They can be left for several years before lifting and replanting in new soil. An annual top dressing of a low-nitrogen granular fertilizer is beneficial in autumn—the nutrients filter down slowly throughout the growing season. *I. danfordiae* is a problem case since large bulbs flower well the first season after purchasing them, but then proceed to split up into many non-flowering bulblets. For regular flowering colonies it is best to plant this species in a bulb frame, the bulbs placed at least 10cm deep. This, for some reason, encourages the individual bulbs to remain at flowering size. This system, incidentally, works very well with other shy-flowering bulbs such as *Crocus sativus*.

I. histrio, I. kolpakowskiana, I. pamphylica and *I. vartanii* all seem best when planted in a bulb frame and I cannot really recommend pot cultivation for any of the reticulata group as they all seem to dwindle away even with annual repotting. For this group the soil, whether it be in the open or in a bulb frame, should be alkaline and I normally mix garden lime in with the compost which is to be used, at the rate of about one handful of lime to a two-gallon bucket of compost—very crude, but I am sure that exact measurements are quite unnecessary. The compost itself is a standard John Innes type of mixture with extra sharp gritty sand added if the loam is rather heavy. As with other Middle East bulbs, they are dried off completely during the dormant season and given fertilizer, as mentioned above, in the autumn. In addition to these species, any special variants of *I. reticulata* are probably best kept in the bulb frame, and *I. bakerana*. *I. histrioides* and *I. winowgradowii*, on the other hand, seem to do much better if planted out in the open ground.

One problem with reticulata irises is Ink Disease, a fungus which appears as black lesions on the surface of the bulb and which can rapidly kill a complete colony. I have had complete success in its control using Benlate, a systemic fungicide. Whenever the bulbs are lifted I dip them in a solution of it or, if in a hurry, directly into the powder. With bulbs which are left in the soil I water a solution of the substance on to the area in the autumn.

The species of Subgenus Hermodactyloides (the 'reticulata irises')

I. bakerana Foster. This is named after J. G. Baker, the famous Kew botanist of the late 19th/early 20th century, who did so much work in the field of bulbous plants. In its typical form from Turkey it is distinct from other species in having nearly cylindrical leaves which have eight ribs running along their length, instead of the usual quadrangular leaves of this group. There is usually a fair amount of leaf visible at flowering time, often overtopping the flower. The flowers are whitish with a deep violet apex to the falls and strong blotches and veining of violet around the creamy-white ridge, and on the haft of the falls. The standards and styles are bluish-lilac. In this Turkish form it is very distinct from all other reticulata irises, but unfortunately in western Iran there are forms which obscure this distinctness. In the border area between Lake Rezaiyeh and the Turkish frontier I have seen plants with the same cylindrical leaves but having smaller pale blue flowers with a yellow crest or ridge, looking exactly like some forms of *I. reticulata*. Farther to the south nearer Shiraz there is one with a strikingly blotched flower more like the Turkish plant, but with four-sided or even five to seven veined leaves. The position is therefore not clear and the true status of *I. bakerana* must be in some doubt until much more field work can be done. It flowers in February or March, growing at about 800–1500 metres in heavy clayey soils on scrub-dotted hillsides in southern Turkey in the Mardin area, in northern Iraq and in western Iran, from Rezaiyeh south to Shiraz.

I. bakerana is easy to grow, but for regular flowering bulb frame cultivation seems to be preferable to the open ground. The commercially available clone, which is probably Turkish in origin, is as easy as *I. reticulata*.

I. bornmülleri Hausskn. A synonym of *I. danfordiae*.

I. danfordiae (Baker) Boissier. A very well-known and distinct little plant, distinguished from all other species in having yellow flowers with reduced bristle-like standards only about 3–5mm long. The only other yellow reticulata group iris is *I. winowgradowii* which has larger, primrose-yellow flowers with larger oblanceolate standards.

I. danfordiae has bulbs which produce many small offset bulblets. The flowers are of a deep yellow with sparse green spotting in the centre and lower part of the falls, and a deeper yellow or orange ridge. The leaves vary greatly in their development at flowering time, from only about 1cm long to just overtopping the flower. It flowers in the wild in March and April and inhabits stony slopes, often flowering near the snow line, and may be in open places or in sparse pine, fir or cedar woods at altitudes of 1000–2000 metres. It is an endemic of Turkey and can be found either in the south, in the eastern end of the Taurus mountains, or in the north around Ordu, Amasya, Sivas Gümüshane and Erzincan. The wild forms I grow are much more attractive than the commercially available clone. They have heather-scented, more richly yellow flowers with wider-spreading falls, but share the annoying habit of producing many tiny bulbs rather than large flowering-sized ones. However, the 'deep planting' method mentioned in connection with the general cultivation of this group seems to work well with *I. danfordiae*.

I. histrio Reichb. fil. This is one of the largest-flowered species in the reticulata group, but also one of the least hardy. The leaves are quadrangular in section and very robust, reaching 30cm or more at flowering time and well-exceeding the flower, and extending to 50–60cm when mature. The flower is about 6–8cm in diameter and is pale blue, fading to almost white towards the centre of the falls and on the haft. The pale area of the blade is covered with large blue blotches and in the central area is a low yellow ridge. The standards and style branches are unspotted. There is a little variation in wild populations and I have seen a form from Lebanon with a darker blue ground colour. *I. histrio* is a native of the Amanus and Taurus Mountains in southern Turkey, south through Syria to Lebanon, inhabiting stony hillsides, often in oak scrub up to 1200 metres, and flowering in January or February.

   I. histrio is a beautiful, very early-flowering species which is not difficult to grow in a cold house or bulb frame, but the bulbs tend to divide up into many tiny bulblets. Deep planting and feeding during the growing season helps to maintain flowering-sized bulbs.

I. histrio var. *aintabensis* G. P. Baker. This is the name given to a small form of *I. histrio* from near Gaziantep in southern Turkey. It was collected for G. P. Baker and a very short note about it appears in the *Gardeners' Chronicle* of 1931. It has usually been treated as a variety of *I. histrio* but is certainly very distinct from the larger-flowered 'normal' form which grows farther south. Var. *aintabensis* has a bulb which produces 'rice-grain' bulblets at its base, and leaves which at flowering time are at most reaching up to the flower and sometimes have the tips only just appearing above soil level. Its flowers are about the size of those of *I. reticulata* and are usually pale blue with some darker splashes of colour on the falls, around the yellow ridge. I have collected the plant myself around Gaziantep and it does vary somewhat in colour to a much deeper blue than the commercially available stock (e.g. Mathew & Tomlinson 4182). Nearer Maraş I collected some dormant bulbs which turned out to be a form in which the falls were tipped with intense dark purple and heavily veined on the haft, with an orange ridge in the centre (Mathew & Tomlinson 4501). In the wild, var. *aintabensis* flowers in March and April in and around corn plots, sometimes at the base of scrub oak or between rocks and it appears to be confined to the Gaziantep-Maraş region of Turkey.

   It will grow in a well-drained soil outside in Britain, but is also very good in a bulb frame.

I. histrio var. *atropurpurea* (Dykes) Dykes is a form which was sent to Dykes from Maraş in Turkey in 1908. He describes it as having a black sheen on a dark red ground and it lacks a yellow central ridge on the falls. As far as I know this is not now in cultivation.

I. histrio var. **orthopetala** Dykes. This is the name Dykes gave to a form with standards which tended to curve in towards each other, and longer bracts which were white with distinct green veins. The origin of it was unknown to him.

I. histrioides (G. F. Wilson) S. Arnott. Of all the reticulatas this is perhaps the best for its large bright flowers, extreme hardiness and ability to maintain itself as flowering-sized bulbs rather than splitting up into tiny bulblets. The quadrangular leaves are nearly absent at flowering time with usually only the tips just showing. Later they develop to up to 40–50cm in length and are wider than those of most other species. The flowers are about 6–7cm in diameter and are variable in depth of blue and amount of spotting in the central area of the blade of the falls. The ridge is rich yellow. *I. histrioides* is apparently confined to a small area around Amasya in central-north Turkey where it flowers in March or April at about 1500 metres on open slopes or in sparse pine woods.

The wild plants which I grow are slightly variable in depth of colour, but mostly resemble the form grown as 'Lady Beatrix Stanley'. There are various cultivar names and in an attempt to define these in 1979 I obtained as many stocks as I could, and checked on the descriptions where available. It seems that there is confusion over the plant grown as 'var. *major*'. This was originally described by Col. Grey in the catalogue of the Hocker Edge Nursery of August 1935 as having 'large light blue flowers in January'. Van Tubergen Ltd were the introducers of this plant from Turkey, where it was collected in the early 20th century by J. J. Manissadjian of Merzifon, near Amasya. In *New Bulbous and Tuberous-rooted Plants* (1947), Van Tubergen Ltd make the following statement. 'We distributed the *Iris histrioides* from Merzifon as *I. histrioides major* as there is also an *I. histrioides* from Cilicia which produces smaller flowers possessing a more violet shade of colour.' It is clear from this statement that the original *major* distributed by Van Tubergen and by Grey was fairly typical of the plant which grows at Amasya and was not in fact larger-flowered, and we know that it had pale blue flowers. The '*I. histrioides* from Cilicia' with smaller flowers was probably a form of *I. histrio*, since I have seen no records of *I. histrioides* from that region at all. To confuse matters, in the last 20 years or so, a plant with very large dark violet-blue flowers with almost horizontal falls has been grown as '*major*'. It could in no way be described as being light blue as in the original description of *major*. The perianth tube of this dark form is stained with dark violet blue. In addition to the original *I. histrioides* var. *major* Grey, which is I believe still distributed correctly by Van Tubergen, and the darker-flowered plant masquerading as *major* there are the cultivars 'G. P. Baker', 'Lady Beatrix Stanley', 'Reine Immaculée' and 'Angel's Tears'.

'G. P. Baker' is undoubtedly the clone figured in tab. 9341 of the *Botanical Magazine*, for there is the statement, 'The material for the present plate was received from Mr. G. P. Baker of Sevenoaks who obtained the bulbs from Amasia.' It is a rather dark blue form, collected in 1909.

'Lady Beatrix Stanley' is paler blue than the above, and heavily spotted on the falls, more so than in var. *major*. It is said to have originated in Lady

Stanley's garden as a variant of var. *major*.

'Reine Immaculée' is a pale, prominently spotted form which was selected and listed by Walter Blom's nursery during the last decade. It is similar to 'Lady Beatrix Stanley'.

'Angel's Tears' is currently offered by the same nursery and they kindly sent me some bulbs for comparison with the other cultivars. They describe it as being a seedling derived from *I. histrioides major* with silky flax-blue flowers, prominently flecked with inky blue, larger and more vigorous than 'Reine Immaculée.'

In addition to these cultivars, Blom's also offer 'Blom's Hybrids' which vary in the depth of colour.

My observations of all these forms are that they are probably all true *I. histrioides* selections with no other species involved, and that the colour differences between them are all very slight. Using the RHS colour chart, I made the following notes:

'major' (Blom's stock): Falls Violet-Blue 93B, shading to 93A around the yellow crest. Standards and styles 93C.

'major' (Van Tubergen stock): Falls Violet-Blue 94B, shading to 94A around the crest. Standards and styles 94B.

'Lady Beatrix Stanley': Falls Violet-Blue 93C, shading to 93B around the crest. Standards and styles 93C.

'Angel's Tears' (Blom): Falls Violet-Blue 94B, shading to 94A around the crest. Standards and styles 94C.

'G. P. Baker': unfortunately this did not flower so I could not compare it in the RHS Chart. It was described in the *Botanical Magazine* in terms of the Ridgway Standard colour chart as 'T9 Smalt Blue'.

'Wild Stock' (from Amasya, collected T. Baytop) showed variation covering a wider range than that of the above forms.

It can be seen that the differences between the various stocks are very slight. Nevertheless, horticulturally speaking, some of the clones are quite distinct and worth maintaining.

I. histrioides var. **sophenensis** (Foster) Dykes. This was originally described by Foster as a variety of *I. reticulata* and it does seem to be rather more like that than *I. histrioides* although its relationships are not very clear and need further investigation. It was collected by Mrs Barnum in 1884 in 'hills near Kharput', which is near the provincial town of Elazığ in Turkey. The plant now grown as this seems to match the original description quite well and similar collections have been made in recent years by Dr Adil Güner of Ankara. The leaves are barely visible at flowering time with only the sharp tips appearing. Since the bulbs produce masses of rice-grain bulblets, there are often many of these leaf tips around each parent bulb. The flowers are deep violet-blue with little veining or spotting and they have a yellow ridge on the falls. The segments are very narrow compared with *I. histrioides* or *I. reticulata*, which gives the flower a certain characteristic appearance and it is certainly distinct as a garden plant whatever its botanical status turns out to be. Some of the collected plants which resemble it have flowers of a more reddish-purple tint than the original.

I. hyrcana Woron. ex Grossheim. For the *Flora Iranica* account of Iris, Per Wendelbo and I included this as a synonym of the widespread and variable *I. reticulata*. Certainly from a botanical point of view there is little to separate the two and it is probably the only course. *I. hyrcana*, named after the Hyrcanian forest region bordering the Caspian Sea, is said to differ in having a nearly spherical bulb producing many bulblets. Plants in cultivation under this name certainly have this characteristic looking bulb, whereas that of *I. reticulata* is longer and pear-drop or almond-shaped. The other noteworthy feature of *I. hyrcana* is the flower colour which is a very clear pale blue with hardly any veining or spotting, and a narrow bright yellow ridge. I have seen and collected reticulata irises in the Talysh region of Iran and they vary in colour from this clear pale blue through darker blue to reddish purple, but always rather clear unbroken colours. The bulbs of this collection (Bowles Scholarship Botanical Expedition 532) are rounded when flowering sized. Unfortunately only the pale blue form of this has survived in cultivation giving the false impression that *I. hyrcana* is always this colour! For garden purposes it is worth distinguishing *I. hyrcana* in some way even if it is very close to *I. reticulata*, so for convenience I shall continue to use the name for the moment. Paul Furse also gathered it in the Caspian region under several numbers, one of which (P.F. 5072) is still in cultivation.

I. kolpakowskiana Regel. This is the black sheep of the reticulata group in that it has three or four narrow, channelled leaves more like those of a small juno iris. However, the yellowish bulb is covered by reticulate tunics and it does not have the fleshy roots of the juno group, so there seems no doubt that it belongs with the reticulatas, albeit a rather divergent species. The leaves are short at flowering time but elongate up to 25cm later. The flower is on a very short stem and has a perianth tube 5–7cm long. There is a little variation in colour in the plants I have seen but usually it is pale lilac-blue or very pale purple with a dark reddish purple blade to the falls. The median ridge on the falls is yellow-orange and the haft is whitish. It is a plant of stony mountain slopes up to 3000 metres altitude in the Tien Shan mountains, flowering in April or May near the melting snow.

Although it has been in cultivation intermittently for over 100 years, *I. kolpakowskiana* has never been more than a rarity in a few specialized collections. It does not seem to be an easy species to grow in Britain and although I grew it and flowered it for several years it has now disappeared. A bulb frame seems to offer the best chance of survival, with a good supply of water in the spring growing season. My plants varied a little in the depth of the background colour of the flower.

I. pamphylica Hedge. This is one of the more recently discovered species, described in 1961 from material gathered by Peter Davis and Oleg Polunin. It is a very distinct reticulata, rather tall when in flower since the flower is carried on a pedicel about 10–20cm long. The quadrangular leaves extend enormously after flowering and can reach 55cm in length. The flowers have rather narrow perianth segments and consist of an extraordinary mixture of colours. The falls have a deep brownish-purple lamina with a bright yellow blotch in the centre which is in turn spotted with purple; the haft is a greenish colour, veined with purple. The standards are light blue, shading to green and flecked with purple-brown in the lower narrow portion; the strongly arching style branches are similarly coloured. There is a little variation in depth of these colours in wild plants. Unlike all other species in the group which have erect capsules, *I. pamphylica* has a capsule which is pendent from a long stalk when it is mature. It inhabits the edge of pine woods and oak scrub at about 700–850 metres in southern Turkey where it flowers in March. It is not an uncommon plant but seems to occur only in a restricted area in Antalya province. In cultivation it responds well to bulb frame treatment but dwindles in size if pot-grown. Although not a large flower, or strikingly coloured, it is a most interesting species apparently not closely related to any other.

I. reticulata M. Bieb. This is the most well-known species in the group, and it is certainly by far the most variable in appearance. The bulbs, when producing offspring, either divide into sizeable bulbs alongside the parent have 'rice-grain' bulblets around its base. The quadrangular leaves vary enormously in their development at flowering time from being scarcely visible to overtopping the flower. In colour, *I. reticulata* is probably at its most variable from the deep violet-blue of the commercial form, through to reddish purples and pale blue. There is usually a yellow crest but sometimes this is lacking. Many garden selections have been made, and there are hybrids as well, some of them with *I. bakerana*. The species is distributed mainly in the Caucasus, northern and eastern Turkey, north-eastern Iraq and Iran, where it is very widespread in the west and north. It occurs in a wide range of habitats and altitudes from 600–2700 metres in open fields, stony mountainsides, alpine meadows or oak scrubland.

The commercial form of *I. reticulata* is very similar to some I have grown from Russian Armenia. In Turkey, the predominant form in the dampish northern mountains is reddish purple with a rather squat flower, appearing almost before the leaves. This is near to the varieties named as *purpurea* and *krelagei*, although there are so many slight variants of the species that it is difficult to supply names with any degree of accuracy. Var. *cyanea* is the name given to a rather small clear blue form; var. *alba* is a beautiful white form (there are some 'whites' which are nearer to a rather washy lilac than white). In Iran most of the forms I have seen are blue, usually with pale blue standards and styles and dark falls with a golden crest.

I. reticulata is easily grown in the open ground in Britain and is a lovely rock-garden plant. It also responds well to slight forcing in bowls indoors, but undoubtedly the best way to grow the wild collected forms is to plant them in a bulb frame.

I. vartanii Foster. This is known in cultivation mainly in its white form which is generally available from specialist bulb nurseries. The quadrangular leaves are usually clearly visible and often overtopping the flower, which is normally blue or slate-coloured in the wild. The characteristics of *I. vartanii* are the very narrow claws to the falls and the long narrow style lobes, which are a very prominent feature of the flower. The crest is cream-coloured. In its method of propagation it is similar to *I. histrio* in producing rice-grain bulblets.

I. vartanii is not a very easy plant in cultivation and tends to dwindle away quite rapidly. It is unfortunate that the white form is the one most readily obtainable, for the blue one I have grown is infinitely more attractive, in a lovely powder-blue shade. The best way to keep the species is in a sunny well-drained bulb frame.

I. vartanii is a native of Israel, and possibly adjacent Syria, and it flowers from November to January at low altitudes in scrub on rocky hillsides.

I. winkleri Regel. A poorly known species, never recollected since it was described in 1884, it is said to have the bulb covered with brown membranous tunics not separating into fibres, and three or four curved leaves which are linear and only 1–2mm wide. The flower is described as being bluish-violet. It was collected in the high mountain zone at 3000–4000 metres in the Fergana area of the Tien Shan mountains and both geographically and morphologically must be close to *I. kolpakowskiana*. The brown membranous tunics would be odd for a member of this group but without some new collections one cannot assess this plant.

I. winowgradowii Fomin. A superb plant, perhaps one of the best of all reticulatas. Its quadrangular leaves are as robust as those of *I. histrioides* which it resembles in general size and habit of growth. The flowers, however, are pale primrose yellow with green spots on the haft of the falls and in the centre of the blade. It is reported from the Adzharo-Imeretian range, on Mount Lomis-mta and from the mountains above Gagre in Abkhazia where it grows in alpine meadows. In cultivation it is an easy species to grow if planted out on the rock garden or sunny bed, but it does not like excessive drying and baking in the summer. Being a mountain plant it is very hardy and likes plenty of moisture in the growing season. The bulb produces plenty of rice-grain bulblets and it is surprising that it is not more commonly seen in gardens. The parent bulbs do not split up into small non-flowering ones as do those of the other yellow-flowered species, *I. danfordiae*.

I. winowgradowii hybridizes with its relative *I. histrioides* and there are now at least two hybrids with intermediate characters. *I.* 'Katharine Hodgkin' is almost certainly a cross between these two, made by E. B. Anderson, and 'Frank Elder' is another named hybrid. Both have flowers with a blue-yellow mixture of colours, and both are hardy vigorous plants for growing outside without protection.

In 1972, Dr Rodionenko expressed the view that *I. winowgradowii* was 'an extremely rare species about to disappear in the wild' and that only a few hundred plants were known to exist. Let us therefore attempt to maintain this beautiful plant in our collections for conservation purposes and not be too anxious to hybridize it out of existence!

Iris narcissiflora

I. narcissiflora Diels. This is a most exciting iris, judging from the herbarium material I have seen. It is of rather unusual appearance and it is difficult to suggest any obvious affinities with other species, especially in the absence of any fruiting specimens to give an idea of the features of the capsules and seeds.

The rootstock apparently consists of slender stolons which are covered with papery scale-leaves, radiating from a small growing point—there is scarcely a trace of a rhizome. There is no basal leaf fan and it seems to me that the plant might in fact produce these separately from the flowering stems. This is not an unusual method of growth of course but it is unfortunate that a sterile leaf fan is not available for study. The flower stem is 20–25cm tall and carries one or two reduced leaves near its base. These are sheathing and have a free tip only about 5cm long. In addition there is one long erect linear leaf about 2mm wide, reaching up to the flower, and the basal 10–15cm of this sheathes the flower stem completely. Both the leaves and the stems are covered with minute rough projections and small sticky glands. The yellow flower is about 6cm in diameter and is subtended by two brown papery ovate bracts 2·5cm long which have very sharply acute tips. One of the striking peculiarities of the flower is its overall roundness, because the falls and standards are nearly equal to each other in size and shape, and each is rounded at the apex. They are narrowed to rather short hafts and joined into a short tube about 6–7mm long. It appears that the flower is flattish, that is with spreading falls and standards, but it is difficult to be sure from dried material. There is a very narrow beard on the falls, only 1–2mm wide, another interesting feature and obviously the cause of its being placed with the bearded species (Section Iris) by some authorities. This I imagine would have been guesswork, without seeing any specimens, for there is no similarity otherwise. The style branches are short, probably only about 1–5cm long.

I. narcissiflora was described as a new species by Diels from material collected in the Szechuan province of China in 1922 at 'Ta-pao-shan'.

The genus *Gynandriris*

This is a small group of iris-like plants consisting of a few species in South Africa and two in the Mediterranean region, extending east to central Asia. The genus can be distinguished from *Iris* by its rootstock being a rather globose or somewhat flattened corm which is covered by a netted tunic. The leaves are narrow and channelled, usually only one or two per corm. There is no perianth tube and the segments are attached directly to the apex of a narrow beak-like extension to the ovary. The ovary develops into a slender capsule containing many very small seeds. In *Iris*, only the 'reticulata' group have netted bulbs but they, like most other iris species, have a distinct perianth tube. A further point about *Gynandriris* is that the flowers, which are produced from transparent papery bracts, are extremely fleeting, each one lasting only a few hours, usually in the afternoon and evening. This is a feature which does not occur to such a marked extent in *Iris* (some may last only a day or two) and this, together with its other characters suggests that *Gynandriris* is more closely related to *Moraea* than to *Iris*.

Cultivation

The northern hemisphere *G. sisyrinchium* is, I find, perfectly hardy but requires a good summer baking in order to make it flower. In countries which do not have predictably warm, dry summers it is therefore best treated as a bulb-frame plant where it can be dried off completely during its dormant period. The same applies to *G. monophylla*.

Those species of the South African winter rainfall (S. W. Cape) area which I have cultivated do well in pots in a frost-free greenhouse. They are started into growth in August or September and dried off when dormancy sets in after producing seeds in the spring. One species *(G. simulans)* is from summer rainfall areas and this should be started into growth about May in Britain. These Cape species are almost certainly capable of being grown outside in Australia, northern New Zealand and California.

Propagation of all the species I have grown is simply a matter of corm division which takes place naturally and quite freely. Seeds are usually easy to obtain and the young corms will flower in about three or four years. The soil I use for all the species is either a John Innes-type mixture or a loamless compost with extra grit added for drainage.

The northern hemisphere *Gynandriris* species

G. sisyrinchium (Linn.) Parl. (Syn. *Iris sisyrinchium*, *I. maricoides*). This is the very common and widespread Barbary Nut, the corm of which is said to have a nut-like flavour. It is very variable, from about 10–40cm in height, normally with two straight, or sometimes arching basal leaves which are channelled and usually about 3–5mm wide. The inflorescence has one to four short erect branches, appearing compact and spike-like. The silvery or transparent bracts give rise to a succession of up to six of the small, 3–4cm diameter, short-lived, bright violet-blue or clear lavender coloured flowers. Either the falls are white in the centre with dark-spotting or they have a yellow signal area. The standards are erect and smaller than the falls and the style branches are also more or less erect and rather pointed. Flowering time is normally in the afternoon and evening, the flowers withering away before morning. The capsules are up to 4cm in length with a persistent narrow beak.

G. *sisyrinchium* occurs throughout the Mediterranean regions from Portugal and North Africa eastwards through Europe, Libya and Egypt to Iran, Afghanistan, Russian Central Asia and Pakistan. It flowers between March and May and may be seen in a range of habitats from open sunny hillsides to cornfields. Occasionally albino forms occur and I have heard that yellowish forms have been seen in Russia, but I cannot confirm this and the *Flora USSR* does not mention such a variant. The very fleeting nature of the flowers makes G. *sisyrinchium* a rather unsatisfactory plant for cultivation, although well-flowered clumps are a delightful asset to a bulb frame after the main display of early spring bulbs is over.

G. monophylla Boiss. & Heldr. ex Klatt. Although described as a species over 100 years ago, G. *monophylla* has been treated by most authorities as a synonym of G. *sisyrinchium*. However, P. Goldblatt has now studied the species again and it is quite clear that it should be regarded as distinct. It is a small plant usually only about 5cm in height and with one narrow leaf (rarely two), prostrate or coiling on the ground. The stem is usually unbranched and produces very small, 2–2·5cm diameter, pale cloudy slate-blue flowers with an orange-yellow signal stripe on the falls. As in G. *sisyrinchium* there is a white zone in the centre of the blade and the haft is spotted.

G. *monophylla* is restricted to the eastern Mediterranean, mainly in open rocky areas near the sea. In Greece it occurs only in Attica and in Crete, and it is also recorded in Cyrenaica and Egypt.

The South African species of Gynandriris

Like G. *sisyrinchium* these all have smallish fleeting flowers in varying shades of blue-violet, or they are white. There are, according to Dr Peter Goldblatt seven species, six of which are natives of the winter rainfall area of the south and south-west Cape and one, G. *simulans*, in the more central summer rainfall region extending into Lesotho, Botswana and Zimbabwe. His revision can be found in *Botaniska Notiser* **133** (1980).

The species are as follows:

G. anomala Goldblatt. Pale blue flowers; signal patch on falls white; height 30cm.

G. australis Goldblatt. Pale blue flowers; signal patch white; height 6–25cm.

G. cedarmontana Goldblatt. White flowers; height 10–30cm.

G. hesperantha Goldblatt. Purple-violet flowers; standards erect; height 40–60cm.

G. pritzeliana (Diels) Goldblatt. Dark blue or violet flowers; standards reflexed; height 12–35cm.

G. setifolia (Linn. f.) Foster. Pale lavender-blue flowers; signal patch yellow; height 5–20cm.

G. simulans (Baker) Foster. Pale blue flowers, speckled all over; height 10–45cm.

The genus Hermodactylus

A curious genus of only one species, obviously closely related to *Iris* but differing from it in having long finger-like tubers and an ovary with only one chamber instead of three as in *Iris*. The leaves are square in cross section, like those of a 'reticulata' iris. *H. tuberosus* is best grown in a warm sunny place, preferably in alkaline or at least neutral soils. It will often grow quite satisfactorily but seldom produces flowers unless given a really hot place such as a south-facing wall can provide. Propagation is by division of the tubers which increase by vegetative means very quickly. They are best moved and divided in August or September just before growth commences.

H. tuberosus (Linn.) Salisbury. The Snake's Head or Widow Iris. The long, narrow, quadrangular leaves overtop the stems, which are usually about 20–40cm in height. There is one flower per stem, about 4–5cm in diameter and rather conical in shape with reflexed velvet brownish-violet blades to the falls. The ground colour is normally a translucent green but yellowish or bronze forms also occur as variants in wild populations. A slender green bract overtops the flower like a hood. The standards are narrowly-oblanceolate and reduced in size to only 2–2·5cm long and the style branches have pointed lobes.

This extraordinarily coloured iris is a common native in Mediterranean regions from south-east France and North Africa eastwards to Israel. It flowers in March or April and is normally found in dry rocky places up to about 1000 metres altitude. The best colony I have ever seen was in the chalky garden of Sir Frederick Stern at Highdown where it was thriving and flowering well on a south-facing bank.

The genus Pardanthopsis

This genus of one species, closely related to and resembling *Iris* in general flower form, is an interesting one although not of great garden value, for the flowers are small and short-lived. Dr Lee W. Lenz has studied and discussed the question (in *Aliso*, Vol. 7, 1972) of the generic status of *Iris dichotoma* (= *Pardanthopsis dichotoma*) and puts a convincing case for recognizing it as belonging to a distinct genus from *Iris*. The most obvious features which distinguish it are the much-branched, many-flowered inflorescence which produces a succession of very short-lived flowers, the lack of a perianth tube and the way in which the flower spirals on shrivelling. Another curious piece of behaviour is that the flower, if unfertilized, immediately breaks off just below the ovary whereas in *Iris* it usually just shrivels away without falling off—if it does so, the break occurs above the ovary. *Pardanthopsis* is more closely related to the genus *Belamcanda* in which the flowers fall off in the same way. A description of the latter can be found on page 186.

The distinctive transverse bands and blotches of colour on the haft of the falls in *Pardanthopsis* is rather reminiscent of the South American genus *Neomarica*, which also has short-lived flowers.

P. dichotoma has been treated taxonomically in several different ways, but always on its own, in a separate Subgenus, Section or Subsection of the genus *Iris*.

P. dichotoma (Pallas) Lenz. An unusual rhizomatous plant which is the only species in its genus. The rhizome is small and has a mass of rather thick roots. The fans of six to eight leaves are nearly as robust as those of a tall bearded iris, when the plant is growing strongly, with each leaf up to 2·5cm wide. The flowering stem reaches 40–100cm in height and is much branched so that there are often many clusters of flowers. Each flower opens in the afternoon and is short-lived but a long succession is produced. They are about 3·5–4·5cm in diameter and variable in colour, although one of the forms I grow which came from seeds collected at the Ming Tombs, Peking is probably the most typical one. The ground colour is a creamy-white with purplish-brown spots and bands on the centre and lower part of the falls. These bands are across the width of the haft, not following the veining as in most, if not all, irises. Another form I have has deep purple flowers with a whitish, purple-spotted patch in the centre of the blade of the falls and the same banding on the haft. Field notes on some specimens say that the flowers

are bluish. The flower is of fairly typical *Iris* shape but the standards are much smaller than the falls, and are rounded at the apex, and the style branches are even smaller, each with two narrow lobes. Unless the flowers are successfully fertilized they wither and fall off immediately, unlike all other irises in which they remain attached to their pedicels for a long period. In my garden I have seen them being visited by hover flies. Capsules are produced fairly readily and these are long and cylindrical, containing many seeds. The species is a native of Mongolia, northern China and Siberia and it occurs in scrub and grassy places up to 2500 metres. In Britain, and in the wild, it flowers in July and August and is an easy plant to grow in full sun with plenty of moisture in summer. Seeds germinate readily and the young plants will flower when only one or two years old. The individual plants are not long-lived and a succession of seedlings should be raised to maintain the stock. *P. dichotoma* has been hybridized by Samuel N. Norris of Kentucky with the related genus *Belamcanda* to produce some extraordinary intermediates, since the latter has regular flowers with six equal perianth segments and the style branches are slender rather than petaloid as in *Pardanthopsis*. The hybrids often resemble *P. dichotoma* in general flower shape but come in varying shades of salmon and apricot, and have reduced style branches. They are known as ×
Pardancanda norrisii Lenz.

The genus Belamcanda

It is generally accepted that this interesting genus contains only one species, although Col. C. N. Grey described and cultivated another at one time, as mentioned below. The one frequently grown species is *B. chinensis*, easily recognized by its flower which has six equal reddish-spotted perianth segments, not differentiated into falls and standards as in an *Iris*. Furthermore, the three styles are slender like those of *Crocus sativus*, with a terminal stigma, not expanded and petaloid like those of irises in which the stigma is a flap on the underside of each of the three style branches. Apart from this, the habit of growth is similar to some irises. *Belamcanda* is undoubtedly most closely related to *Pardanthopsis dichotoma (Iris dichotoma)* and will hybridize with it to produce × *Pardancanda* hybrids which I have mentioned under the latter genus.

In cultivation in Britain *Belamcanda* presents no problems if given reasonably good soil with plenty of humus in sun or semi-shade. It does not like a very warm dry position and should have plenty of moisture in the growing season. I find that it is completely hardy in Surrey but is not a long-lived plant. It is however easily raised from seed and flowers in two or three years from sowing.

B. chinensis (Linn.) DC. (Syn. *B. punctata*). The small, often somewhat stoloniferous rhizomes, give rise to fans of broad leaves, each about 1–2cm wide, and leafy stems which reach 60–100cm in height at flowering time. The inflorescence is widely branched with about three to twelve flowers about 4cm in diameter. These have six equal perianth segments which are a yellowish or orange-red colour mottled with red or blackish-purple spots. They have hardly any perianth tube at all and the pedicels are jointed just below the ovary so that the whole flower quickly falls off from this point if it is not fertilized. The three style branches are slender, not petaloid. Unlike irises, the capsules split open and the three locules curl outwards leaving the central axis exposed. The large blackish seeds stay attached for a considerable time before falling, this feature having given rise to the common name of Blackberry Lily. *Belamcanda chinensis* is a native of Japan, China, eastern Russia in the Ussuri region, Taiwan and northern India. It occurs in sandy meadows near the sea, in moist scrubland and in shady places from sea level to about 2000 metres altitude.

B. flabellata Grey. In his *Hardy Bulbs*, Col. Grey writes that this plant was collected in Japan, but I can find no other record of it in literature, or herbarium specimens. He describes it as being about 30–45cm tall and having the stems covered with closely-packed leaves. The eight to fourteen flowers are carried in a lax inflorescence and are yellow, spotted with orange-yellow near to the base of the segments. He recommends that it is grown in shade with plenty of moisture. I have not heard of this plant in cultivation and it may well be lost.

Glossary

Acid   A soil which has a pH of less than 7; opposite of alkaline.

Acuminate   Long-pointed

Acute   Pointed

Aggregate   In botany usually used for a group of closely related 'species' which are difficult to distinguish from each other.

Albino   A white variant of a flower which is normally coloured.

Alkaline   Soil with a pH of more than 7; opposite of acid, usually chalky or limestone

Appendage   An extra attachment to an organ, often of no apparent use.

Aril   An attachment to the seed, usually fleshy in fresh seeds. It is an expanded part of the funicle arising from the placenta.

Axil   The junction between leaf and stem.

Axillary   Growing from an axil.

Basal   Often applied to a leaf or leaves which arise at ground level, not on the stem.

Beard   In *Iris*, an area of hairs in the centre and lower part of the falls (outer perianth segments) and sometimes also the standards.

Bifid   Divided into two.

Blade   The expanded portion of a leaf or perianth segment. In *Iris* refers to the larger flattened portion of the falls and standards.

Bract   A modified leaf subtending the pedicel of a flower, often enclosing it in bud. (cf. spathe).

Bracteole   A secondary bract, often smaller.

Bulb   Underground storage organ consisting of fleshy scales attached to a basal plate of solid tissue and enclosing a growing point.

Bulblet   Small bulbs produced around the parent bulb.

Calcareous   Growing in chalky or limestone soils.

Capsule   A dry seed pod which splits lengthways in *Iris* to shed its seeds.

Ciliate   Fringed with hairs.

Clone   A plant which is propagated vegetatively so that all the individuals are genetically identical.

Colony   A group of individuals of a species.

Concolorous   Of a uniform colour.

Corm   Underground storage organ consisting of solid tissue, not scaly like a bulb and normally replaced by a new one each year.

Crest   In *Iris* referring to the ridge, which is sometimes dissected, in the centre of the falls. Frequently coloured yellow.

Cultivar   A variant of a plant, considered to be distinct from a horticultural point of view and maintained in cultivation. Nowadays must be given non-latinized names.

Cylindric   Tube-like, with a circular cross section.

Deflexed   Bent downwards, as in the standards of Scorpiris (juno) species.

Dilated   Swollen or expanded.

Distichous   Flattened into one plane, as in the 'fan' of leaves of bearded irises.

Divergent   Moving away from each other; held apart. In the Californicae group of irises the tips of the bracts in some species are divergent. In others they are held together.

Endemic   In botany refers to a plant which is confined to a given area such as one country, one mountain, an island etc.

Entire   Undivided, untoothed. When applied to falls or standards of an iris, also means that they have no lobes.

Falcate   Curved in a sickle-like manner.

Falls   The outer perianth segments of an iris which usually 'fall' downwards at their tips.

Filament   The stalk of a stamen.

Filiform   Very narrow and thread-like.

Fruit   Any mature seed-bearing organ, whatever its form.

Genus (pl. genera)   A group of related plants all bearing the same generic name (e.g. *Iris*), subdivided into species each with a separate specific name (e.g. *Iris sibirica, Iris reticulata* etc.)

Glabrous   Smooth, without any hairs or teeth.

Gland   An organ which secretes a sticky substance; often hair-like.

Glaucous   Covered with a greyish waxy coat, like cabbage leaves.

Habitat   The conditions encountered in the locality in which a plant grows.

Haft   The narrow basal portion of the falls or standards of an iris.

Herbaceous   Often meaning green, not transparent, when referring to bracts.

Herbarium   A collection of dried plant specimens.

Hybrid   The result of a cross between two or more different plants, whether they be species, subspecies, varieties etc.

Inflorescence   The whole flowering portion of a plant consisting of stem, bracts and flowers.

Internode   The piece of stem between two joints (nodes) at which leaves or bracts arise.

Keel   Ridged along the central vein as in the keel of a boat; in *Iris* applied to the bracts, which may have a sharp keel.

Lamina   The expanded portion of a leaf or perianth segment.

Lanceolate   Tapering at both ends but broadest just below the middle.

Lateral   Arising from the side, such as a side-branch of a flower stem.

Lax   Loose, spaced out.

Linear   Narrow with the edges parallel.

Linear-lanceolate   Very narrowly lanceolate, bordering on linear.

Lobe   A projection of a leaf or perianth segment; in *Iris* the standards of Scorpiris species are sometimes three-lobed.

Local   Referring to distribution, a species which is somewhat restricted, but not necessarily very rare.

Locules   The chambers of a capsule; in *Iris* there are three.

Membranous   Thin and semi-transparent.

Mesophytic   Plants which occur in conditions which are neither extremely dry (xerophytic) or extremely wet (aquatic).

Monograph   A written account of one particular group of plants.

Naturalized   Of foreign origin but established and reproducing as if a native.

Nectary   An organ in which nectar is secreted, usually near the base of a flower.

Oblanceolate   Tapering at both ends but broadest just above the middle.

Oblong   Longer than broad, with more or less parallel sides.

Obovate   Reversed egg-shaped, broadest near the apex.

Obtuse   Blunt at the apex.

Offset   A small vegetatively produced bulb at the base of the parent bulb.

Opposite   Arising from the same node, one on each side of the stem, usually referring to leaves or branches of the inflorescence.

Orbicular   Having a circular outline.

Ovary   The female portion of the flower containing the ovules which after fertilization develop into the seeds.

Ovate   Egg-shaped.

Papillose   Covered with minute hair-like protuberances.

Pedicel   The stalk of a single flower.

Peduncle   The main stalk of the whole inflorescence, each flower then being carried on a pedicel.

Perianth   The outer, usually showy part of a flower; in *Iris* refers to the three falls and three standards, (i.e. six perianth segments).

Perianth tube   The part of a flower where the perianth segments join together to form a tube; not always present, as in *Gynandriris*.

Pubescent   Hairy.

Recurved   Curved outwards and downwards.

Reflexed   Abruptly bent downwards.

Reticulate   Netted.

Rhizome   The root-like, often swollen, stock of various plants (e.g. *Iris*) which is capable of producing both roots and shoots. It has a terminal bud and lateral ones in the axes of leaves or scales.

Rootstock   A term usually applied to the basal part of a plant whether it be a corm, bulb, rhizome etc.

Scarious   Dry and membranous, not green.

Segment   One of the divisions of a perianth, compound leaf etc.

Sessile   Stemless, having no pedicel, the flower carried directly on the peduncle which may also be nearly absent so that the flower is carried at ground level.

Spathe   A modified leaf, often much reduced and papery, enclosing the inflorescence in bud.

Species   A unit of classification; a group of individuals which have a common set of characters.

Spike   A raceme-like inflorescence in which the individual flowers are stemless on the elongated axis.

Stamen   The male part of a flower, consisting of anther and filament, producing pollen.

Standard   In *Iris*, the three inner perianth segments which usually 'stand up' erect, but in Scorpiris (juno) they are reflexed.

Sterile   Usually applied to flowers which are incapable of producing seeds because of some deformity or aberration in morphology, cytology etc.

Stigma   The tip of the female part of the flower which receives the pollen. In *Iris*, a flap on the underside of the style branch.

Stolon   In *Iris* refers to an underground stem produced by a rhizome which gives rise to further plants; it is like a slender extension of the rhizome.

Style   The 'stem' of the female portion of the flower, between the ovary and the stigma. In *Iris* there are three style branches and each is flattened and petal-like, and bi-lobed at its apex.

Subspecies   A unit of classification below that of species.

Synonyms   Different names which refer to the same species.

Taxon   An unspecified unit of classification.

Taxonomy   The study of classification.

Terminal   Produced at the apex, usually referring to a flower or inflorescence.

Terra-Rossa   The red clay soil which is very common in much of the Mediterranean region.

Toothed   Applied to a margin of leaf, perianth segment etc. which is jagged with teeth-like projections.

Tuber   A swollen subterranean organ; some *Iris* of the Nepalensis group have tuber-like roots.

Tunic   The outer coat of bulbs or corms.

Type   The original specimen from which the description of a species was made.

Type locality   The place in which the type specimen was collected.

Unilocular   Applied to a capsule which has only one chamber, not three as in *Iris*.

Variety   A unit subordinate to species and subspecies.
Widespread   Distributed over a wide area, but not necessarily common in any one place.
Xerophytic   Growing in very dry situations.

Selected bibliography

American Iris Society, (1972), *Species Manual*, ed. B. L. Davidson, Seattle
American Iris Society, annual *Bulletins*
Awishai, M., (1977), *Species relationships and cytogenetic affinities in Section Oncocyclus of the genus Iris* (Doctorate thesis), Hebrew University, Jerusalem

Boissier, E., (1882), *Flora Orientalis*, Vo. 5, Geneva
British Iris Society (Species Group), *Summary of Iris Species*, London
British Iris Society, *Iris Year Book*, London

Chaudhary, S. A., (1975), 'Iris subgenus Susiana in Lebanon and Syria' in *Botaniska Notiser* **128**, Lund, Sweden
Cohen, V. A., (1967), *A Guide to the Pacific Coast Irises*, British Iris Society, London
Curtis' Botanical Magazine, (1787-   ), London

Davis, P. H., (1946), 'Oncocyclus Irises in the Levant' in *Journal of the Royal Horticultural Society*, **71**, London.
Dykes, W. R. (1913), *The Genus Iris*, Cambridge
Dykes, W. R., (1924), *Handbook of Garden Irises*, London

Fedtshenko & Vvedensky, 'Iris' in Vol. 4 of *Flora USSR* ed. V. L. Komarov, Leningrad; English ed. 1968
Foster, M., *Bulbous Irises*, London (1892), Royal Horticultural Society
Foster., R. C., 'A cytotaxonomic survey of the North American species of *Iris*' in *Contributions from the Gray Herbarium* no. 119 (1937), Cambridge Mass., USA

Gardeners' Chronicle, many articles on *Iris*, from 1841 onwards, London
Goldblatt, P., 'Systematics of Gynandriris' in *Botaniska Notiser* **133** (1980), Lund, Sweden
Grey, C., *Hardy Bulbs*, 3 vols, (1938), New York
Grey-Wilson, C., *The Genus Iris Subsection Sibiricae*, the British Irish Society (1971), London
Grossheim, A. A., 'Iris' in Vol. 2 of *Flora Kavkaza* (Caucasus) (1940), Leningrad
Guner, A., *Turkish Iris* (Doctorate thesis) (1979), Ankara

Index Kewensis, 4 vols. + 15 supplements (1905–1970 continuing), Oxford University Press
International Rules of Botanical Nomenclature, latest edition Utrecht (1978)

Journal of Royal Horticultural Society, many articles on *Iris* (1846–   ), London

Kew Record of Taxonomic Literture, annual record of all new works (1971–   ), London
Lawrence, G. H. M., 'A reclassification of the genus *Iris*' in *Gentes Herbarium* **8**:346 (1953), New York
Lee, Y. N., 'New Taxa on Korean Flora' in *Korean Journal of Botany* **17**, no. 1 (1974), Seoul
Lee, Y. N., *Illustrated Flora and Fauna of Korea*, Vol. 18 (1976), Seoul
Lenz, L. W., 'A revision of the Pacific Coast Irises', in *Aliso* **4**:1 (1958), California
Lenz, L. W., 'Hybridization and speciation in the Pacific Coast Irises' in *Aliso* **4**:237 (1959), California
Lenz, L. W., '*Iris tenuis*' in *Aliso* **4**:311 (1959), California
Lenz, L. W. 'A key character in *Iris* for separating the Sibiricae and the Californicae' in *Aliso* **5**:211 (1962), California

192   Selected bibliography

Lenz, L. W. 'The status of *Pardanthopsis*' in *Aliso* **7**:401 (1972), California
Lenz, L. W., 'Studies in *Iris* embryo culture' in *Aliso* **3**:173 (1955), California
Lenz, L. W., and Day, A., 'The chromosomes of the Spuria irises and evolution of the garden forms' in *Aliso* **5**:257 (1963), California
Lynch, R. I., *The Book of the Iris* (1904), London and New York

Maire, R., *'Iris'* in Vol. **6** of *Flore de L'Afrique du Nord* (1959), Paris
Marchant, A., & Mathew, B., *An alphabetical table and cultivation guide to the species of the genus* Iris, the British Iris Society (1974), London
Mathew, B., an account of *Iridaceae* (unpublished) in *Flora of Iraq*, ed. C. C. Townsend & E. Guest, Iraq Ministry of Agriculture, Baghdad
Mathew, B., an account of *Iris* (unpublished) in *Flora of Turkey*, ed. P. H. Davis, University Press, Edinburgh
Mouterde, P., *Nouvelle Flora du Liban et de la Syrie*, Vol. 1 (1966), Beirut
Murashige, T., 'Plant propagation through tissue cultures' in *Ann. Rev. Plant Physiol.* **25**:135 (1974)

Ohwi, J., *Flora of Japan* (1965), Shibundo, Tokyo, translated edition, Smithsonian Institution, Washington DC

Post, G., *Flora of Syria, Palestine and Sinai* (1896), Beirut, ed. 2 by J. Dinsmore, (1933), Oxford University Press
Prodan, J., 'Die Iris—Arten Rumaniens' in *Bul. Grad. Bot. Univ. Cluj.* **14** (1934); **15** (1935); **25** (1946), Cluj, Roumania
Prodan, J., *'Iris'* in Vol. IIAX OF *Flora of Roumania* (1966), Bucharest

Randolph, L. F. (ed.), *Garden Irises* (1959), American Iris Society
Randolph, L. F., *'Iris nelsonii'* in *Baileya* **14**:143 (1966), New York
Randolph, L. F. and F., 'Embryo culture of *Iris* seed' in *Bul. Amer. Iris Soc.* **139**:7 (1955)
Randolph, L. F., & Mitra, J., 'Karyotypes of *Iris* pumila and related species' in *Amer. Journ. Bot.* **46**:93 (1959), Baltimore
Report of First International Symposium on Iris, the Societa Italiana dell'Iris (1963), Florence
Rodionenko, G. I., *Rod (Genus)* Iris (1961), Leningrad

Small, J. K. & Alexander, E. J., 'Botanical Interpretations of the Iridaceous plants of the Gulf States' in *Contr. New York Bot. Gard.* **327**:325 (1931), New York
Stojanov, N., Stefanov, B., & Kitanov, B., 'Iris' in part I, 4th ed. of *Flora of Bulgaria* (1966)

Taylor, John J. 'A reclassification of *Iris* species bearing arillate seeds' in *Proc. Biol. Soc. Washington* **89**, 35:411 (1976), Washington

Ugrinsky, K. A., 'The *Iris flavissima* Pallas complex' in *Fedde Repert. Spec. Nov.* **14** (1922), Berlin

Vvedensky, A. I., *Conspectus Florae Asiae Mediae* Vol. 2 (1971)

Warburton, B., & Hamblen, M. (eds), *The World of Irises*, American Iris Society (1978), Wichita, Kansas
Webb, D. A., & Chater, A. O., 'Iris' in Vol. 5 of *Flora Europaea* (1980), Cambridge
Wendelbo, P., & Mathew, B., 'Iridaceae' in *Flora Iranica*, part 112 (1976), Akademische Druck-v. Verlagsanstalt, Graz, Austria
Werckmeister, P., *Catalogus Iridis* (1967), Deutsche Iris—und Lilien gesellschaft

Index to Iris species

Figures in **bold type** denote main reference to that species. Figures in *italic* indicate a line drawing on that page